A BEGINNER'S FURTHER GUIDE
TO MATHEMATICAL LOGIC

A BEGINNER'S FURTHER GUIDE
TO MATHEMATICAL LOGIC

Raymond Smullyan

World Scientific

NEW JERSEY · LONDON · SINGAPORE · BEIJING · SHANGHAI · HONG KONG · TAIPEI · CHENNAI · TOKYO

Published by

World Scientific Publishing Co. Pte. Ltd.

5 Toh Tuck Link, Singapore 596224

USA office: 27 Warren Street, Suite 401-402, Hackensack, NJ 07601

UK office: 57 Shelton Street, Covent Garden, London WC2H 9HE

Library of Congress Cataloging-in-Publication Data

Names: Smullyan, Raymond M., author.

Title: A beginner's further guide to mathematical logic / Raymond Smullyan.

Description: New Jersey : World Scientific, 2016. |
   Includes bibliographical references and index.

Identifiers: LCCN 2015033651 | ISBN 9789814730990 (hardcover : alk. paper) |
   ISBN 9789814725729 (pbk : alk. paper)

Subjects: LCSH: Logic, Symbolic and mathematical.

Classification: LCC QA9.A1 S619 2016 | DDC 511.3--dc23

LC record available at http://lccn.loc.gov/2015033651

British Library Cataloguing-in-Publication Data

A catalogue record for this book is available from the British Library.

On the cover, the three photos from left to right are the logicians
Emil Post, Alan Turing, and Ernst Zermelo.

Typeset by Stallion Press

Email: enquiries@stallionpress.com

Printed in Singapore

Contents

Part II: Recursion Theory and Metamathematics   67

Part III: Elements of Combinatory Logic   171

Preface

This book is a sequel to my *Beginner's Guide to Mathematical Logic* [Smullyan, 2014]. I originally intended both volumes to be a single volume, but I felt that at my age (now 96), I could pass away at any time, and I wanted to be sure that I would at least get the basic material out.

The previous volume deals with elements of propositional and first-order logic, contains a bit on formal systems and recursion, and concludes with chapters on Gödel's famous incompleteness theorem, along with related results.

The present volume begins with a bit more on propositional and first-order logic, followed by what I would call a "fein" chapter, which simultaneously generalizes some results from recursion theory, first-order arithmetic systems, and what I dub a "decision machine." Then come four chapters on formal systems, recursion theory and metamathematical applications in a general setting. The concluding five chapters are on the beautiful subject of combinatory logic, which is not only intriguing in its own right, but has important applications to computer science. Argonne National Laboratory is especially involved in these applications, and I am proud to say that its members have found use for some of my results in combinatory logic.

This book does not cover such important subjects as set theory, model theory, proof theory, and modern developments in recursion theory, but the reader, after studying this volume, will be amply prepared for the study of these more advanced topics.

Although this book is written for beginners, there are two chapters — namely 3 and 8 — that I believe would also be of interest to the expert.

For brevity, all references to the first volume, *The Beginner's Guide to Mathematical Logic*, of this two-volume introduction to mathematical logic will be given in the remainder of this volume as *The Beginner's Guide* [Smullyan, 2014].

Elka Park

November 2016

Part I

More on Propositional and First-Order Logic

Chapter 1

More on Propositional Logic

I. Propositional Logic and the Boolean Algebra of Sets

Many of you have probably noticed the similarity of the logical connectives to the Boolean operations on sets. Indeed, for any two sets A and B and any element x, the element x belongs to the intersection $A \cap B$ if and only if x is in A and x is in B. Thus $x \in (A \cap B)$ iff $(x \in A) \wedge (x \in B)$. Thus the logical connective $\wedge$ (conjunction) corresponds to the Boolean operation $\cap$ (intersection). Likewise the logical connective $\vee$ (disjunction) corresponds to the Boolean operation $\cup$ (union), since $x \in (A \cup B)$ iff $(x \in A) \vee (x \in B)$ Also, $x \in \overline{A}$ (x is in the complement of A) if and only if $\sim (x \in A)$ (x is *not* in A), so that the logical connective negation corresponds to Boolean operation of complementation.

Note: As in *The Beginner's Guide* [Smullyan, 2014], I will often abbreviate the phrase "if and only if" by "iff", following the valuable suggestion of Paul Halmos.

We saw in Chapter 1 of *The Beginner's Guide* how to verify the correctness of a Boolean equation by the method of what I called "indexing". However, due to the correspondence between the logical connectives and the Boolean operations on sets, one can also verify Boolean equations by truth tables. The following single example will illustrate the general idea. Consider the Boolean equation $\overline{A \cap B} = \overline{A} \cup \overline{B}$. This equation is valid iff for every element x, the element x is in $\overline{A \cap B}$ iff x is in $\overline{A} \cup \overline{B}$. Thus the equation $\overline{A \cap B} = \overline{A} \cup \overline{B}$ is to the effect that for all x,

$$x \in \overline{A \cap B} \text{ iff } x \in \overline{A} \cup \overline{B}. \text{ And}$$
$$x \in A \cap B \text{ iff } (x \in A) \wedge (x \in B). \text{ Hence,}$$
$$x \in \overline{A \cap B} \text{ iff } \sim((x \in A) \wedge (x \in B)). \text{ Similarly,}$$
$$x \in \overline{A} \cup \overline{B} \text{ iff } \sim(x \in A) \vee \sim(x \in B).$$

Thus the proposition $x \in \overline{A \cap B}$ iff $x \in (\overline{A} \cup \overline{B})$ is equivalent to

$$\sim((x \in A) \wedge (x \in B)) \equiv \sim(x \in A) \vee \sim(x \in B).$$

And that formula is a tautology, for it is an instance of $\sim(p \wedge q) \equiv \sim p \vee \sim q$, as can be seen by replacing $(x \in A)$ by p and replacing $(x \in B)$ by q.

Similarly, the Boolean equation $A \cap (B \cup C) = (A \cap B) \cup (A \cap C)$ holds since $p \wedge (q \vee r) \equiv (p \wedge q) \vee (p \wedge r)$ is a tautology.

In general, given a Boolean equation, define its *counterpart* to be the result of substituting $\wedge$ for $\cap$ and $\vee$ for $\cup$; $\sim t$ for $\bar{t}$ (where t is any set variable or Boolean term), $\equiv$ for $=$, and propositional variables p, q, r, etc., for set variables A, B, C, etc. For example, the counterpart of the Boolean equation $\overline{A} \cup B = \overline{A \cap \overline{B}}$ is the propositional formula $\sim p \vee q \equiv \sim(p \wedge \sim q)$.

Well, a Boolean equation is valid if and only if its counterpart is a tautology of propositional logic. And since the formula $\sim p \vee q \equiv \sim(p \wedge \sim q)$ is again easily seen to be a tautology, the Boolean equation $\overline{A} \cup B = \overline{A \cap \overline{B}}$ is seen to be valid.

Thus, truth tables can be used to test Boolean equations. Going in the other direction, the method of "indexing," described in Chapter 1 of *The Beginner's Guide* [Smullyan, 2014] can be used to verify a statement of propositional logic, after making the appropriate Boolean equation counterpart to the propositional statement (found by reversing the substitutions mentioned above).

II.   An Algebraic Approach

This approach to propositional logic is particularly interesting and leads to the subject known as *Boolean rings*.

For any two natural numbers x and y, if x and y are both even, so is their product $x \times y$. Thus an even times an even is even. An even times an odd is even. An odd times an even is even, but an odd times an odd is odd. Let us summarize this in the following table.

x	y	$x \times y$
E	E	E
E	O	E
O	E	E
O	O	O

Note the similarity of this table to the truth table for disjunction. Indeed, if we replace E by T, O by F, and $\times$ by $\vee$, we get the truth table for disjunction:

x	y	$x \vee y$
T	T	T
T	F	T
F	T	T
F	F	F

This is not surprising, since x times y is even if and only if x is even *OR* y is even.

However, this might well suggest (as it did to me, many years ago!) that we might transfer the problem of whether a given formula is a tautology, to the question of whether or not a certain algebraic expression in variables for natural numbers is always even (for all possible values of the variables). So let us now do this, i.e. define a transformation of a formula of propositional logic into an arithmetic expression (using only $+$ and $\times$ and variables for natural numbers) such that the propositional formula will be a tautology iff every number corresponding to the numerical value of an instance of the arithmetic expression (i.e. an instance in which natural numbers replace the variables) is even.

We let *even* correspond to *truth*, *odd* to *falsehood*, and *multiplication* to *disjunction*; then we take the letters $p, q, r, \ldots$ both as variables for *propositions* in logical expressions and as variables for *natural numbers* in the corresponding arithmetic expressions. Thus the logical expression $p \vee q$ is now transformed to the arithmetic expression $p \times q$, or more simply, pq.

What should we take for *negation*? Well, just as $\sim$ transforms truth into falsehood, and falsehood into truth, we want an operation that transforms an even number to an odd one, and an odd one to an even one. The obvious choice is to add 1. Thus we let the logical expression $\sim p$ correspond with the numerical expression $p+1$.

As was seen in Chapter 5 of *The Beginner's Guide* [Smullyan, 2014], once we have $\vee$ and $\sim$, we can obtain all the other logical connectives. We take $p \supset q$ (*implication*) to be $\sim p \vee q$, which transforms to $(p+1)q$, which is $pq+q$.

What about *conjunction*? Well, we take $p \wedge q$ to be $\sim(\sim p \vee \sim q)$, and so in the numerical interpretation we take $p \wedge q$ to be $((p+1)(q+1))+1$, which is $(pq+p+q+1)+1$, which is $pq+p+q+2$. But 2 is an even number, which means that for any number z, the number $z+2$ has the same parity as z (i.e. $z+2$ is even if and only if z is even). And so we can drop the final 2, and take $p \wedge q$ to be $pq+p+q$.

We could take $p \equiv q$ to be $(p \supset q) \wedge (q \supset p)$ or we could take $p \equiv q$ to be $((p \wedge q) \vee (\sim p \wedge \sim q))$. But it is much simpler to do the following: Now, $p \equiv q$ is true iff p and q are either both true or both false. Thus in our numerical interpretation, we want the number $p \equiv q$ to be even if and only if p and q are either both even or both odd. And the simple operation that does the trick is addition: we thus take $p \equiv q$ to be $p+q$.

We note that $p \vee p$ is equivalent to p and thus $p \times p$ has the same parity as p. So we can replace $p \times p$ (or pp) by p. Also, for any number p, $p+p$ is always even, and so we can replace $p+p$ with 0 (which is always even). Moreover, we can eliminate any *even number* that occurs as an *addend*, because eliminating an even addend does not change the parity of a numerical expression: e.g. $p+0$ and $p+4$ is of the same parity as p, so can be replaced by p.

Under the transformations just described, a formula is a tautology in the usual propositional interpretation, if and only if it is always even in its numerical interpretation.

Let us consider the tautology $(p \supset q) \equiv \sim (p \wedge \sim q)$. Well, $p \supset q$ corresponds to $pq+q$. Now, $p \wedge \sim q$ transforms to $p(q+1)+p+(q+1)$, which simplifies to $pq+p+p+q+1$. We can drop $p+p$, which is always even, and so the transformation of $p \wedge \sim q$ reduces to $pq+q+1$. Thus $\sim(p \wedge \sim q)$ reduces to $pq+q+1+1$, but $1+1$ is even, and can be dropped, so that $\sim(p \wedge \sim q)$ reduces to $pq+q$, which is the same thing that $p \supset q$ reduced to. Thus $(p \supset q) \equiv \sim(p \wedge \sim q)$ reduces to $(pq+q)+(pq+q)$, which expression can be reduced to 0 (since, as mentioned above, the sum of the same two equal numbers is always even). We have thus shown that $(p \supset q) \equiv \sim(p \wedge \sim q)$ is a tautology by showing that the corresponding numerical expression it reduces to (in this case 0) is always even.

Note: Readers familiar with modern higher algebra know that a Boolean ring is a ring with the additional properties that $x + x = 0$ and $x^2 = x$ for any element x of the ring.

III. Another Completeness Proof

We now turn to a completeness proof for propositional logic of an entirely different nature, and one which has a particularly interesting extension to first-order logic, which we consider in the next chapter. It is based on the notion of *maximal consistency*.

We will consider axiom systems using $\sim, \wedge, \vee,$ and $\supset$ as undefined connectives, and in which modus ponens (from X and $X \supset Y$, to infer Y) is the only rule of inference. We will say that a *Boolean truth set* is a set S of formulas such that for any pair of formulas X and Y, the following conditions hold:

(1) $(X \wedge Y) \in S$ iff $X \in S$ and $Y \in S$.
(2) $(X \vee Y) \in S$ iff either $X \in S$ or $Y \in S$.
(3) $(X \supset Y) \in S$ iff either $X \notin S$ or $Y \in S$.
(4) $(\sim X) \in S$ iff $X \notin S$ (thus, one and only one of $X, \sim X$ is in S).

Notes: [a] Whenever necessary to avoid a possible misreading, given a formula X and a set S, I will abbreviate the statement that $\sim X$ is a member of S by $(\sim X) \in S$ instead of by $\sim X \in S$, as the latter could be read as "It is not the case that X is in S." Similarly, I write $(\sim X) \notin S$ rather than $\sim X \notin S$ to mean that $\sim X$ is not in S, for $\sim X \notin S$ might be read as $\sim(X \notin S)$, i.e. that it is not the case that X is not a member of S (so that X *is* in S). Outer parentheses around formulas are often omitted for easier reading, but may be included for greater clarity of meaning; for instance, parentheses are used to distinguish between $(X \wedge Y) \in S$ from $X \wedge (Y \in S)$, two possible readings of $X \wedge Y \in S$ (where X and Y are variables for sentences, and S is a variable representing a set of sentences). [b] The reader should also realize that for any statement of the form "X iff Y", the statement is equivalent to its negated *form* "$\sim X$ iff $\sim Y$", by the very meaning of the term "iff" (which is an abbreviation for "if and only if"). Thus,

$$\text{“}(\sim X) \in S \text{ iff } X \notin S\text{”}$$

is equivalent to

$$\text{“}(\sim X) \notin S \text{ iff } X \in S\text{”}.$$

If I refer to an assumption or condition in a proof or a problem solution, but want the reader to take it in its equivalent "negated" form, I often indicate this fact.

We also recall from *The Beginner's Guide* [Smullyan, 2014] the unifying α, β notation in propositional logic for unsigned formulas:

α	α_1	α_2
$X \wedge Y$	X	Y
$\sim(X \vee Y)$	$\sim X$	$\sim Y$
$\sim(X \supset Y)$	X	$\sim Y$
$\sim\sim X$	X	X

β	β_1	β_2
$\sim(X \wedge Y)$	$\sim X$	$\sim Y$
$X \vee Y$	X	Y
$X \supset Y$	$\sim X$	Y

Problem 1. S is a Boolean truth set if and only if for every α and every formula X the following conditions hold:

(1) $\alpha \in S$ iff $\alpha_1 \in S$ and $\alpha_2 \in S$.
(2) Either $X \in S$ or $(\sim X) \in S$, but not both.

Exercise 1. Prove that S is a Boolean truth set iff for every β and every formula X the following conditions hold:

(1) $\beta \in S$ iff either $\beta_1 \in S$ or $\beta_2 \in S$.
(2) Either $X \in S$ or $(\sim X) \in S$, but not both.

The definition of a Boolean truth set here can be shown to be equivalent to the one given for a *truth set* in Chapter 6 (Propositional Tableaux) of *The Beginner's Guide* [Smullyan, 2014], where we called a set S of propositional formulas a *truth set* if it was the set of all formulas true under some interpretation I of the propositional variables. We also showed there that the definition was equivalent to the set S satisfying the following three conditions: for any signed formula X and any α and β:

T_0: Either X or its conjugate $\overline{X}$ is in S, but not both.
T_1: α is in S if and only if α_1 and α_2 are both in S.
T_2: β is in S if and only if either β_1 is in S or β_2 is in S.

With these facts in mind, the reader should easily see that the definitions of (Boolean) truth set in *The Beginner's Guide* and this work are equivalent. Note that once one sees the equivalence between the definitions in the two books, it is obvious that every Boolean truth set for propositional logic is satisfiable (since the characterization in *The Beginner's Guide* made a truth set satisfiable *by definition*).

Indeed, to go a little further (to make connections between very similar concepts a little clearer), we often define a *Boolean valuation* as follows. First, by a *valuation* (in the context of propositional logic) is meant an assignment of truth values t and f to all formulas. A valuation v is called a *Boolean valuation* if for all formulas X and Y, the following conditions hold:

B_1: $\sim X$ receives the value t under v if and only if $v(X) = f$ [i.e. when X is false under v]; and accordingly $v(\sim X) = f$ iff $v(X) = t$.

B_2: $X \wedge Y$ receives the value t under v iff $v(X) = v(Y) = t$ [i.e. when both X and Y are true under the valuation v].

B_3: $X \vee Y$ receives the value t under v iff $v(X) = t$ or $v(Y) = t$ [i.e. when either X is true under v or Y is true under v, or both are true under v].

B_4: $X \supset Y$ receives the value t under v iff $v(X) = f$ or $v(Y) = t$ [i.e. when either X is false under v or Y is true under v].

The *interpretations* that we used in *The Beginner's Guide* in terms of assigning truth values to all formulas by assigning truth values to the propositional variables, and then working up through the structure of the formulas, using the meanings of the connectives, to obtain a truth value for every formula, are closely related to the truth values of formulas given through a Boolean valuation. In fact a given Boolean valuation defines a unique interpretation yielding the same truth value on all formulas in the interpretation as in the Boolean valuation: we obtain the interpretation by looking at the truth values the Boolean valuation assigns to the propositional variables. And if one is given an interpretation of the propositional variables, it is clear that the interpretation defines a Boolean valuation on the formulas, just by the way the interpretation assigns truth values to all formulas, starting out only with truth values on the propositional variables.

Because of this tight equivalence between Boolean valuations and interpretations in propositional logic, we will use the words "valuation" and "interpretation" interchangeably (and will do so later in first-order logic as well, where a similar tight relationship holds, this being something that was already discussed in *The Beginner's Guide* [Smullyan, 2014]).

Consequence Relations

We consider a relation $S \vdash X$ between a set S of formulas and a formula X. We read $S \vdash X$ as "X is a consequence of S," or as "S yields X".

We call the relation $\vdash$ a *consequence relation* if for all sets S, S_1 and S_2 and all formulas X and Y, the following conditions hold:

C_1: If $X \in S$, then $S \vdash X$.
C_2: If $S_1 \vdash X$ and $S_1 \subseteq S_2$ then $S_2 \vdash X$.
C_3: If $S \vdash X$ and $S, X \vdash Y$, then $S \vdash Y$.
C_4: If $S \vdash X$, then $F \vdash X$ for some finite subset F of S.

In what follows, we assume that $\vdash$ is a consequence relation. We write $\vdash X$ to mean $\emptyset \vdash X$, where $\emptyset$ is the empty set. For any n, we write $X_1, \ldots, X_n \vdash Y$ to mean $\{X_1, \ldots, X_n\} \vdash Y$, and for any set S, we write $S, X_1, \ldots, X_n$ to mean $S \cup \{X_1, \ldots, X_n\}$.

An important example of a consequence relation is this. For any axiom system $\mathcal{A}$, take $S \vdash X$ to mean that X is derivable from the elements of S together with the axioms of $\mathcal{A}$ by use of the inference rules of the system; in other words, if we take the elements of S and add them as new axioms of $\mathcal{A}$, then X becomes provable in that enlarged system. Note that under this relation $\vdash$, the statement $\vdash X$ simply says that X is provable in the system $\mathcal{A}$. It is obvious that this relation $\vdash$ satisfies conditions C_1, C_2, and C_3. As for C_4, if X is derivable from S, any derivation uses only finitely many elements of S.

We now return to consequence relations in general.

Problem 2. Prove that if $S \vdash X$ and $S \vdash Y$, and $X, Y \vdash Z$, then $S \vdash Z$.

Consistency

In what follows, "set" shall mean a set of formulas.

We shall call a set S *inconsistent* (with respect to a consequence relationship $\vdash$) if $S \vdash Z$ for every formula Z; otherwise, we shall call S *consistent*.

Problem 3. Show that if S is consistent, and $S \vdash X$, then S, X is consistent.

Problem 4. Suppose S is consistent. Show the following:
(a) If $S \vdash X$ and $X \vdash Y$, then S, Y is consistent.
(b) If $S \vdash X$ and $S \vdash Y$ and $X, Y \vdash Z$, then S, Z is consistent.

Maximal Consistency

We say that S is *maximally consistent* (with respect to $\vdash$) if no *proper* superset of S is consistent. [By a proper superset of S is meant a set S' such that $S \subseteq S'$ and S' contains one or more elements not in S.]

Problem 5. Show that if M is maximally consistent, then for all formulas X, Y and Z:
(a) If $X \in M$ and $X \vdash Y$, then $Y \in M$.
(b) If $X \in M$ and $Y \in M$ and $X, Y \vdash Z$ then $Z \in M$.

Boolean Consequence Relations

We shall define a consequence relation $\vdash$ to be a *Boolean consequence relation* if it satisfies the following additional conditions (for all formulas α, X and Y):
C_5: (a) $\alpha \vdash \alpha_1$; (b) $\alpha \vdash \alpha_2$.
C_6: $\alpha_1, \alpha_2 \vdash \alpha$.
C_7: $X, {\sim}X \vdash Y$.
C_8: If $S, {\sim}X \vdash X$, then $S \vdash X$.

We say that an interpretation *satisfies* a set S if all elements of S are true under the interpretation. We shall say that X is a *tautological consequence* of S if X is true under all interpretations that satisfy S.

Exercise 2. Define $S \models X$ to mean that X is a tautological consequence of S. Prove that this relation $\models$ is a Boolean consequence relation. (**Hint:** For C_4, use the Corollary to the Compactness Theorem for Propositional Logic on p. 95 in *The Beginner's Guide* [Smullyan, 2014].)

We say a consequence relation $\vdash$ is *tautologically complete* if, for every set S and formula X, whenever X is a tautological consequence of S, then $S \vdash X$.

Problem 6. Suppose ⊢ is tautologically complete. Does it necessarily follow that $\emptyset \vdash X$ for every tautology X? [We recall that $\emptyset$ is the empty set.]

The following, which we aim to prove, will be our main result:

Theorem 1. *Every Boolean consequence relation is tautologically complete. Thus, if ⊢ is a Boolean consequence relation, then:*
(1) *If X is a tautological consequence of S, then $S \vdash X$.*
(2) *If X is a tautology, then ⊢ X (i.e. $\emptyset \vdash X$). This means that every system of propositional logic based on a Boolean consequence relation is also complete (i.e. that all tautologies are provable within the system).*

To prove Theorem 1, we will need the following lemma:

Lemma 1. *[Key Lemma] If ⊢ is a Boolean consequence relation and M is maximally consistent with respect to ⊢, then M is a Boolean truth set.*

Problem 7. Prove Lemma 1.

Compactness

We recall from Chapter 4 of *The Beginner's Guide* [Smullyan, 2014] that a property P of sets is called *compact* if for any set S, S has the property P if and only if all *finite* subsets of S have property P. Thus, to say that consistency is compact is to say that a set S is consistent if and only if all finite subsets of S are consistent.

Problem 8. [A key fact!] Show that for a Boolean consequence relation ⊢, consistency (with respect to ⊢) is compact.

Let us also recall the *denumerable compactness theorem* of Chapter 4 of *The Beginner's Guide*, which says that for any compact property P of subsets of a denumerable set A, any subset of A having property P can be extended to (i.e. is a subset of) a maximal subset of A having property P. We have just shown that for a Boolean consequence relation, consistency is a compact property. Hence, any consistent set is a subset of a maximally consistent set. This fact, together with the Key Lemma, easily yields Theorem 1.

Problem 9. Now prove Theorem 1.

IV. Fidelity to Modus Ponens

We shall say that a consequence relation $\vdash$ is *faithful to modus ponens* if

$$X, X \supset Y \vdash Y$$

holds for all X and Y.

In what follows, we shall assume that $\vdash$ is faithful to modus ponens. We shall say that a formula X is *provable* with respect to $\vdash$ when $\vdash X$ holds (i.e. if $\emptyset \vdash X$). Until further notice, "provable" will mean provable with respect to $\vdash$.

Problem 10. Show that if $S \vdash X$ and $S \vdash X \supset Y$, then $S \vdash Y$.

Problem 11. Show the following:
(a) If $X \supset Y$ is provable, and $S \vdash X$, then $S \vdash Y$.
(b) If $X \supset (Y \supset Z)$ is provable, and $S \vdash X$ and $S \vdash Y$, then $S \vdash Z$.
(c) If $X \supset Y$ is provable, then $X \vdash Y$.
(d) If $X \supset (Y \supset Z)$ is provable, then $X, Y \vdash Z$.

We now consider an axiom system $\mathcal{A}$ for propositional logic in which modus ponens is the only inference rule. Such a system I will call a *standard system*. By a *deduction* from a set S of formulas is meant a finite sequence $X_1, \ldots, X_n$ of formulas, such that for each term X_i of the sequence, X_i is either an axiom (of $\mathcal{A}$), or a member of S, or is derivable from two earlier terms by one application of modus ponens. We call a deduction $X_1, \ldots, X_n$ from S a deduction of X from S if $X_n = X$. We say that X is deducible from S to mean that there is a deduction of X from S. We now take $S \vdash X$ to be the relation that X is deducible from S. This relation is indeed a consequence relation.

Problem 12. Prove that $\vdash$ is a consequence relation.

It is obvious that this relation $\vdash$ is faithful to modus ponens, since the sequence $X, X \supset Y, Y$ is a deduction of Y from $\{X, X \supset Y\}$.

We say that the axiom system $\mathcal{A}$ has the *deduction property* if for all sets S and formulas X and Y, if $S, X \vdash Y$, then $S \vdash X \supset Y$.

Theorem 2. [The Deduction Theorem] A sufficient condition for an axiom system $\mathcal{A}$ to have the deduction property is that the following are provable in $\mathcal{A}$ (for all X, Y and Z):

A_1: $X \supset (Y \supset X)$.

A_2: $(X \supset (Y \supset Z)) \supset ((X \supset Y) \supset (X \supset Z))$.

For this, we need:

Lemma 2. If A_1 and A_2 hold, then $X \supset X$ is provable.

Until further notice, we assume that A_1 and A_2 hold.

Problem 13. To prove the above lemma, show that for any formulas X and Y, the formula $(X \supset Y) \supset (X \supset X)$ is provable. Then, taking $Y \supset X$ for Y in the formula just proved, conclude the proof of Lemma 2.

Problem 14. In preparation for the proof of the Deduction Theorem, show the following:

(a) If $S \vdash Y$, then $S \vdash X \supset Y$.

(b) If $S \vdash X \supset (Z \supset Y)$ and $S \vdash X \supset Z$, then $S \vdash X \supset Y$.

Problem 15. Prove the Deduction Theorem as follows: Suppose $S, X \vdash Y$. Then there is a deduction $Y_1, \ldots, Y_n$ from $S \cup \{X\}$ in which $Y = Y_n$. Now consider the sequence $X \supset Y_1, \ldots, X \supset Y_n$. Show that for all $i \leq n$, if $S \vdash X \supset Y_j$ for all $j < i$, then $S \vdash X \supset Y_i$. [**Hint:** Break the proof up into the following four cases: Y_i is an axiom of $\mathcal{A}$; $Y_i \in S$; $Y_i = X$; Y_i comes from two terms of the sequence $Y_1, \ldots, Y_{i-1}$ by modus ponens.]

It then follows by complete mathematical induction that $S \vdash X \supset Y_i$ for all $i \leq n$, and so in particular, $S \vdash X \supset Y_n$, and thus $S \vdash X \supset Y$. [**Hint:** In the original sequence $Y_1, \ldots, Y_n$, the term Y_i is either an axiom of $\mathcal{A}$, or is X itself, or comes from two earlier terms by modus ponens. Handle each case separately.]

We now have all the parts necessary to prove:

Theorem 3. A sufficient condition for a standard axiom system for propositional logic $\mathcal{A}$ to be tautologically complete is that for all X, Y, Z and α, the following are provable:

A_1: $X \supset (Y \supset X)$

A_2: $(X \supset (Y \supset Z)) \supset ((X \supset Y) \supset (X \supset Z))$

A_3: (a) $\alpha \supset \alpha_1$ and (b) $\alpha \supset \alpha_2$

A_4: $\alpha_1 \supset (\alpha_2 \supset \alpha)$

A_5: $X \supset (\sim X \supset Y)$
A_6: $(\sim X \supset X) \supset X$

Problem 16. Prove Theorem 3 by showing that under the conditions given in Theorem 3, the relation $\vdash$ is a Boolean consequence relation and thus tautologically complete.

One can easily verify that $A_1 - A_6$ are all provable in the axiom system S of Chapter 7 of *The Beginner's Guide* [Smullyan, 2014]. [One can make a tableau proof for each of these formulas and then convert them to proofs in S by the method explained in that chapter. The completeness proof for tableaux is not needed for this.] We thus have a second proof of the completeness of S based on maximal consistency, and thus the theorem:

Theorem 4. *The axiom system S of Chapter 7 of The Beginner's Guide is tautologically complete.*

Maximal consistency and Lindenbaum's Lemma also play a key role in alternative completeness proofs for first-order logic, to which we turn in the next chapter.

Solutions to the Problems of Chapter 1

1. We are to show that S is a Boolean truth set iff the two conditions of Problem 1 hold. The second condition of Problem 1 is the same as the fourth condition in the definition of the Boolean truth set, so we need only consider the first condition of Problem 1. (Remember, when thinking about "negated" conditions or assumptions, that $\sim(X \wedge Y)$ is equivalent to $\sim X \vee \sim Y$ and $\sim(X \vee Y)$ is equivalent to $\sim X \wedge \sim Y$.)

   (a) Suppose S is a Boolean truth set. We must show that in all cases for α, the first condition of Problem 1 holds, namely that $\alpha \in S$ iff $\alpha_1 \in S$ and $\alpha_2 \in S$.

   Case 1. α is of the form $X \wedge Y$. This is immediate.

   Case 2. α is of the form $\sim(X \vee Y)$. Then $\alpha_1 = \sim X$ and $\alpha_2 = \sim Y$. Thus we are to show that $(\sim(X \vee Y)) \in S$ iff $(\sim X) \in S$ and $(\sim Y) \in S$. Well,

   - $(\sim(X \vee Y)) \in S$ is true iff $(X \vee Y) \notin S$ [by (4) in the definition of a Boolean truth set].

- Then $(X \vee Y) \notin S$ is true iff $X \notin S$ and $Y \notin S$ [by the negated form of (2) in the definition of a Boolean truth set].
- And $X \notin S$ and $Y \notin S$ is true iff $(\sim X) \in S$ and $(\sim Y) \in S$ [by (4) in the definition of a Boolean truth set].

Case 3. α is of the form $\sim(X \supset Y)$. Then $\alpha_1 = X$ and $\alpha_2 = \sim Y$. We are to show that $(\sim(X \supset Y)) \in S$ iff $X \in S$ and $(\sim Y) \in S$. Well,

- $(\sim(X \supset Y)) \in S$ is true iff $X \supset Y \notin S$ [by (4) in the definition of a Boolean truth set].
- Then $X \supset Y \notin S$ is true iff $X \in S$ and $Y \notin S$ [by the negated form of (3) in the definition of a Boolean truth set].
- And $X \in S$ and $Y \notin S$ is true iff $X \in S$ and $(\sim Y) \in S$ [by (4) in the definition of a Boolean truth set].

Case 4. α is of the form $\sim\sim X$. Here both α_1 and α_2 are X, so we must show that $(\sim\sim X) \in S$ iff $X \in S$. Well, $(\sim\sim X) \in S$ iff $(\sim X) \notin S$ [by (4) in the definition of a Boolean truth set], which is true iff $X \in S$ [again by (4) in the definition of a Boolean truth set].

(b) Going in the other direction, suppose the conditions of Problem 1 hold: (i) $\alpha \in S$ iff $\alpha_1 \in S$ and $\alpha_2 \in S$, and (ii) for any X, either $X \in S$ or $\sim X \in S$, but not both. We are to show that S is a Boolean truth set.

Case 1. We are to show that $X \wedge Y \in S$ iff $X \in S$ and $Y \in S$: Well $X \wedge Y$ is an α, where $X = \alpha_1$ and $Y = \alpha_2$ and $\alpha \in S$ iff $\alpha_1 \in S$ and $\alpha_2 \in S$ [by assumption (i) above]. Thus it is immediate that $X \wedge Y \in S$ iff $X \in S$ and $Y \in S$.

Case 2. We are to show that $X \vee Y \in S$ iff $X \in S$ or $Y \in S$: Note that here $\sim(X \vee Y)$ is an α, with $\alpha_1 = \sim X$ and $\alpha_2 = \sim Y$. Well,

- $X \vee Y \in S$ iff $\sim(X \vee Y) \notin S$ [by assumption (ii) above].
- $\sim(X \vee Y) \notin S$ is true iff $(\sim X) \notin S$ or $(\sim Y) \notin S$ [by the negated form of assumption (i)].
- Then $(\sim X) \notin S$ iff $X \in S$ and $(\sim Y) \notin S$ is true $Y \in S$ [by assumption (ii) above].

Case 3. We are to show that $(X \supset Y) \in S$ iff $X \notin S$ or $Y \in S$. Note that here $\sim(X \supset Y)$ is an α, with $\alpha_1 = X$ and $\alpha_2 = \sim Y$. Well,

- $(X \supset Y) \in S$ iff $\sim(X \supset Y) \notin S$ [by assumption (ii)].

- Then $\sim(X \supset Y) \notin S$ is true iff $X \notin S$ or $(\sim Y) \notin S$ [by the negated form of assumption (i) above].
- Since it is true that $(\sim Y) \notin S$ iff $Y \in S$ [by assumption (ii) above], we now have $X \notin S$ or $Y \in S$, which is what was to be shown.

Case 4. To show that $(\sim X) \in S$ iff $X \notin S$: As noted in the definition of a Boolean truth set, condition (4) of a truth set and condition (2) of Problem 1 are equivalent.

2. Suppose $S \vdash X$ and $S \vdash Y$ and $X, Y \vdash Z$. Since $X, Y \vdash Z$, then $S, X, Y \vdash Z$ [by C_2, since $\{X, Y\} \subseteq S \cup \{X, Y\}$]. Since $S \vdash Y$, then $S, X \vdash Y$ [again by C_2]. Thus $S, X \vdash Y$ and $S, X, Y \vdash Z$. Hence [by C_3], $S, X \vdash Z$. [To see this more easily, let $W = S \cup \{X\}$.] Then $W \vdash Y$ (since this is the same as $S, X \vdash Y$). And $W, Y \vdash Z$ (since this is the same as $S, X, Y \vdash Z$). Hence $W \vdash Z$, by C_3, taking W for S. Then, since $S, X \vdash Z$ and $S \vdash X$ (given), then $S \vdash Z$ [again by C_3].

3. Suppose $S \vdash X$. We are to show that if S is consistent, so is S, X (i.e. so is the set $S \cup \{X\}$). We will show the equivalent fact that if S, X is inconsistent, so is S. Well, suppose S, X is inconsistent. Thus $S, X \vdash Z$ for every Z. Since also $S \vdash X$ (which is given), then $S \vdash Z$ [by C_3], and thus S is also inconsistent.

4. Suppose S is consistent.

   (a) Assume $S \vdash X$ and $X \vdash Y$. We are to show that S, Y is consistent. Well, since $X \vdash Y$, then $S, X \vdash Y$ [by C_2]. Since also $S \vdash X$, we have $S \vdash Y$ [by C_3]. Since S is consistent and $S \vdash Y$, then S, Y is consistent [by Problem 3].

   (b) Assume $S \vdash X$ and $S \vdash Y$ and $X, Y \vdash Z$. We are to show that S, Z is consistent. Well, since $S \vdash X$ and $S \vdash Y$ and $X, Y \vdash Z$, it must also be true that $S \vdash Z$ [by Problem 2]. Then, since S is consistent and $S \vdash Z$, we have that S, Z is consistent [by Problem 3].

5. Suppose M is maximally consistent. We first note that if $M \vdash X$, then $X \in M$. For suppose $M \vdash X$. Since M is consistent, then M, X is consistent [by Problem 3]. Consequently M, X cannot be a proper superset of M (no consistent set is). Thus $X \in M$. Now we consider the two parts of the problem.

   (a) Suppose $X \in M$ and $X \vdash Y$. Since $X \in M$, then $M \vdash X$ [by C_1]. Since $X \vdash Y$ and $M \vdash X$, we obtain $M \vdash Y$ [by C_2]. Hence, as just noted, $Y \in M$.

(b) Suppose $X \in M$ and $Y \in M$ and $X, Y \vdash Z$. Then $M \vdash X$, $M \vdash Y$ [by C_1] and $X, Y \vdash Z$. Thus $M \vdash Z$ [by Problem 2]. Hence $Z \in M$, as noted above.

6. Yes, if $\vdash$ is tautologically complete, then $\emptyset \vdash X$ for every tautology X. Here is the proof. Since there are no elements in the empty set $\emptyset$, it is vacuously true that every element of the empty set is true under all interpretations. (If you don't believe this, just try to find an element of the empty set which is not true under all interpretations!) Thus *every* interpretation satisfies both $\emptyset$ and X, so X is a tautological consequence of $\emptyset$. Then, since $\vdash$ is tautologically complete, it follows from the definition of a consequence relation being tautologically complete that $\emptyset \vdash X$ (in fact $S \vdash X$ holds for every set S [by C_2]).

7. We are given that $\vdash$ is a Boolean consequence relation and that M is maximally consistent with respect to $\vdash$. By virtue of Problem 1, to show that M is a Boolean truth set, it suffices to show that the following two things hold for every α and every X:

(1) $\alpha \in M$ if and only if $\alpha_1 \in M$ and $\alpha_2 \in M$.

(2) Either $X \in M$ or $(\sim X) \in M$, but not both.

Regarding (1): Suppose $\alpha \in M$. Since $\alpha \vdash \alpha_1$ [by $C_5(a)$], then $\alpha_1 \in M$ [by Problem 5(a)]. Thus if $\alpha \in M$, then $\alpha_1 \in M$. Similarly, if $\alpha \in M$, then $\alpha_2 \in M$ [by $C_5(b)$]. Thus if $\alpha \in M$, then $\alpha_1 \in M$ and $\alpha_2 \in M$.

Conversely, suppose $\alpha_1 \in M$ and $\alpha_2 \in M$. We know that $\alpha_1, \alpha_2 \vdash \alpha$ [by C_6]. Hence $\alpha \in M$ [by Problem 5(b)].

Thus $\alpha \in M$ if and only if $\alpha_1 \in M$ and $\alpha_2 \in M$.

Regarding (2): Since M is consistent, it cannot be that X and $\sim X$ are both in M, for if they were, then we would have both $M \vdash X$ and $M \vdash \sim X$, and since $X, \sim X \vdash Z$ for every Z [by C_7], we would have $M \vdash Z$ [by Problem 2], which would mean that M is inconsistent. Thus, since M is consistent, it cannot be that $X \in M$ and $\sim X \in M$. It remains to show that either $X \in M$ or $(\sim X) \in M$. Let me show that if $\sim X$ is not in M, then $X \in M$.

Well, suppose that $\sim X$ is not in M. Then $M, \sim X$ is not consistent (since M is maximally consistent). Hence $M, \sim X \vdash X$ (since M, $\sim X \vdash Z$ for every Z), and therefore $M \vdash X$ [by C_8]. Thus $X \in M$ (since M is maximally consistent).

This completes the proof.

8. To show that if all finite subsets of a set S are consistent, then so is S, it suffices to show that if S is inconsistent, then some finite subset of S is inconsistent. Well, suppose S is inconsistent. Then for any formula X, we have both $S \vdash X$ and $S \vdash {\sim}X$ (since $S \vdash Z$ for every Z). Since $S \vdash X$, then $F_1 \vdash X$ for some finite subset F_1 of S [by C_4]. Similarly, $F_2 \vdash {\sim}X$ for some finite subset F_2 of S. Let $F = F_1 \cup F_2$. Then $F \vdash X$ and $F \vdash {\sim}X$ [by C_2], since $F_1 \subseteq F$ and $F_2 \subseteq F$). Also, $X, {\sim}X \vdash Z$ for all Z [by C_7]. Thus, for all Z, $F \vdash Z$ [by Problem 2], and we have shown that F is inconsistent.

9. We wish to show that every Boolean consequence relation is tautologically complete. By Problem 8 we know that for any Boolean consequence relation, consistency with respect to that relation is a compact property, and so [by the denumerable compactness theorem] every consistent set is a subset of a maximally consistent set, which must be a Boolean truth set [by the Key Lemma]. Thus every consistent set is satisfiable (by the discussion after the definition of a Boolean truth set), and hence every unsatisfiable set is inconsistent. Now, suppose X is a logical consequence of S. Then $S, {\sim}X$ is unsatisfiable, hence inconsistent. Therefore $S, {\sim}X \vdash X$ (since $S, {\sim}X \vdash Z$ for every Z), and so $S \vdash X$ [by C_8]. This concludes the proof.

10. Suppose $S \vdash X$ and $S \vdash X \supset Y$. It is also true that $X, X \supset Y \vdash Y$, since we assume $\vdash$ is faithful to modus ponens. Hence $S \vdash Y$ [by Problem 2].

11. Proofs of the four statements:
    (a) Suppose $X \supset Y$ is provable and $S \vdash X$. Since $X \supset Y$ is provable ($\emptyset \vdash X \supset Y$), then $S \vdash X \supset Y$ [by C_2]. Thus $S \vdash X$ and $S \vdash X \supset Y$; hence $S \vdash Y$ [by Problem 10].
    (b) Suppose $X \supset (Y \supset Z)$ is provable, and $S \vdash X$ and $S \vdash Y$. Since $S \vdash X$ and $S \vdash X \supset (Y \supset Z)$ [by C_2], then $S \vdash Y \supset Z$ [by Problem 10]. Since also $S \vdash Y$, then $S \vdash Z$ [again by Problem 10].
    (c) In 11(a), we take the unit set $\{X\}$ for S, and thus if $X \supset Y$ is provable and $X \vdash X$, then $X \vdash Y$. But $X \vdash X$ does hold, so that if $X \supset Y$ is provable, then $X \vdash Y$.
    (d) In 11(b), we take the set $\{X, Y\}$ for S, and thus if $X \supset (Y \supset Z)$ is provable, and $X, Y \vdash X$, and $X, Y \vdash Y$, then $X, Y \vdash Z$. But $X, Y \vdash X$ and $X, Y \vdash Y$ both hold, and so if $X \supset (Y \supset Z)$ is provable, then $X, Y \vdash Z$.

12. We must verify that conditions C_1–C_4 hold.

C_1: If $X \in S$, then the unit sequence X (i.e. the sequence whose only term is X) is a deduction of X from S.

C_2: Suppose $S_1 \subseteq S_2$. Then any deduction $Y_1, \ldots, Y_n$ from S_1 is also a deduction from S_2.

C_3: If $Y_1, \ldots, Y_n$ is a deduction of X from S, and $Z_1, \ldots, Z_k$ is a deduction of Y from S, X, then the sequence $Y_1, \ldots, Y_n, Z_1, \ldots, Z_k$ is a deduction of Y from S (as is easily verified).

C_4: Any deduction from S uses only finitely many elements $X_1, \ldots, X_n$ of S, hence is also a deduction from the set $X_1, \ldots, X_n$.

13. (a) We assume A_1 and A_2 hold. In A_2 we take X for Z, and obtain

$$\vdash (X \supset (Y \supset X)) \supset ((X \supset Y) \supset (X \supset X)).$$

But by A_1, we have $\vdash X \supset (Y \supset X)$. Hence, by modus ponens, we have

$$\vdash (X \supset Y) \supset (X \supset X).$$

(b) In $(X \supset Y) \supset (X \supset X)$, taking $(Y \supset X)$ for Y, we have

$$\vdash (X \supset (Y \supset X)) \supset (X \supset X).$$

But by A_1 we have that $X \supset (Y \supset X)$ is provable. So, by modus ponens, we also have $\vdash X \supset X$.

14. (a) Suppose $S \vdash Y$. Now, by (a) of Problem 11, for any formulas W_1 and W_2, if $S \vdash W_1$ and $W_1 \supset W_2$ is provable, then $S \vdash W_2$. Let us take Y for W_1 and $X \supset Y$ for W_2. When we do so, we see that if $S \vdash Y$ and $Y \supset (X \supset Y)$ is provable, then $S \vdash X \supset Y$. However, $S \vdash Y$ [by assumption] and $Y \supset (X \supset Y)$ is provable [by A_1]; hence $S \vdash X \supset Y$.

(b) By Problem 11(b), for any formulas W_1, W_2, and W_3, if $W_1 \supset (W_2 \supset W_3)$ is provable, then if $S \vdash W_1$ and $S \vdash W_2$, it follows that $S \vdash W_3$. We now take $X \supset (Z \supset Y)$ for W_1, $X \supset Z$ for W_2, and $X \supset Y$ for W_3. Well, the formula $W_1 \supset (W_2 \supset W_3)$ is provable [by A_2; it is the formula $(X \supset (Z \supset Y)) \supset ((X \supset Z) \supset (X \supset Y))$]. Hence, if $S \vdash W_1$ and $S \vdash W_2$, then $S \vdash W_3$, i.e. we have that if $S \vdash X \supset (Z \supset Y)$ and $S \vdash X \supset Z$, then $S \vdash X \supset Y$, which is what was to be proved.

15. We assume that for all $j < i$, $S \vdash X \supset Y_j$, and we are to prove that $S \vdash X \supset Y_i$.

Cases 1 and 2. Suppose that either Y_i is an axiom of $\mathcal{A}$, or that $Y_i \in S$. If the former, then Y_i is provable, so that $\vdash Y_i$, and so $S \vdash Y_i$ [by C_2]. If $Y_i \in S$, then again $S \vdash Y_i$ [by C_1]. Thus, in either case, $S \vdash Y_i$. Hence $S \vdash X \supset Y_i$, [by Problem 14(a)].

Case 3. $Y_i = X$. Since $X \supset X$ is provable [by Lemma 2], then $S \vdash X \supset X$ [by C_2], and since $Y_i = X$, then $X \supset Y_i$ is the formula $X \supset X$, and so $S \vdash X \supset Y_i$.

Case 4. This is the case where in the original sequence $Y_1, \ldots, Y_n$, the formula Y_i came from terms earlier than Y_i by modus ponens. Thus there is a formula Z such that Z and $Z \supset Y_i$ are earlier terms than Y_i. Hence in the new sequence $X \supset Y_1, \ldots, X \supset Y_n$ the formulas $X \supset Z$ and $X \supset (Z \supset Y_i)$ are earlier terms than $X \supset Y_i$, and so, by the inductive hypothesis, $S \vdash X \supset Z$ and $S \vdash X \supset (Z \supset Y_i)$. Then by Problem 14(b), $S \vdash X \supset Y_i$. This completes the proof.

16. We already know that conditions $C_1 - C_4$ hold, and so it remains to verify conditions $C_5 - C_8$.

    Re C_5: (a) Since $\alpha \supset \alpha_1$ is provable [by $A_3(a)$], then $\alpha \vdash \alpha_1$ [by Problem 11(c)].

    (b) Since $\alpha \supset \alpha_2$ is provable [by $A_3(b)$], then $\alpha \vdash \alpha_2$ [by Problem 11(c)].

    Re C_6: Since $\alpha_1 \supset (\alpha_2 \supset \alpha)$ is provable [by A_4], then $\alpha_1, \alpha_2 \vdash \alpha$ [by Problem 11(d)].

    Re C_7: Since $X \supset (\sim X \supset Y)$ is provable [by A_5], then $X, \sim X \vdash Y$ [by Problem 11(d)].

    Re C_8: [This is the only part that needs the Deduction Theorem!] Suppose that $S, \sim X \vdash X$ Then, by the deduction theorem, $S \vdash \sim X \supset X$. But also $(\sim X \supset X) \supset X$ is provable [by A_6]. Therefore $S \vdash X$ [by Problem 11(a), taking $\sim X \supset X$ for X and X for Y].

    This concludes the proof that $\vdash$ is a Boolean consequence relation, and so by Theorem 1, it follows that all tautologies are provable in the system (and more generally, that if X is a logical consequence of S, then $S \vdash X$ holds).

Chapter 2

More on First-Order Logic

I. Magic Sets

There is another approach to Completeness Theorems, the Skolem–Löwenheim Theorem and the Regularity Theorem that has an almost magic-like simplicity. I accordingly dub the central player a *magic set*.

As in propositional logic, in the context of first-order logic a *valuation* is an assignment of truth values t and f to all sentences (with or without parameters). And a valuation v is called a *Boolean valuation* if for all sentences (closed formulas) X and Y, the following conditions hold:

B_1: The sentence $\sim X$ receives the value t under v if and only if $v(X) = f$ [i.e. when X is false under v]; and accordingly $v(\sim X) = f$ iff $v(X) = t$.

B_2: The sentence $X \wedge Y$ receives the value t under v iff $v(X) = v(Y) = t$ [i.e. when both X and Y are true under the valuation v].

B_3: The sentence $X \vee Y$ receives the value t under v iff $v(X) = t$ or $v(Y) = t$ [i.e. when either X is true under v or Y is true under v].

B_4: The sentence $X \supset Y$ receives the value t under v iff $v(X) = f$ or $v(Y) = t$ [i.e. when either X is false under v or Y is true under v].

By a *first-order valuation* (in the denumerable domain of the parameters) is meant a Boolean valuation satisfying the following two additional conditions:

F_1: The sentence $\forall x \varphi(x)$ is true under v if and only if for every parameter a, the sentence $\varphi(a)$ is true under v.

F_2: The sentence $\exists x \varphi(x)$ is true under v if and only if the sentence $\varphi(a)$ is true under v for at least one parameter a.

23

Let S be a set of sentences of first-order logic. We say that S is *truth-functionally satisfiable*, if there is a Boolean valuation in which every sentence in S is true. Similarly, we say that S is *first-order satisfiable* (in first-order contexts, just *satisfiable*), if there is a first-order valuation in which every sentence in S is true.

Note: Recall that it was noted in *The Beginner's Guide* [Smullyan, 2014] that if we consider the set of all true sentences under a first-order valuation v, and look at the universe in which the individuals to which predicates apply as consisting of all the parameters, we can define an interpretation on that universe in which each n-ary predicate is true precisely on the n-tuples of parameters on which the valuation v says it is true (and false on the rest of n-tuples of parameters). This is called an *interpretation in the domain of the parameters*, and because of this we can say a set of sentences S is satisfiable *in the denumerable domain of the parameters* whenever there is a first-order valuation that satisfies S. Although here we are not using the concept of an interpretation of first-order logic as defined in *The Beginner's Guide*, it is easy to see that any interpretation of first-order logic as defined there simultaneously defines a first-order valuation as defined here.

We leave it to the reader to verify the following:

Lemma. *From F_1 and F_2 it follows that under any first-order valuation,*

(1) $\sim\exists x\varphi(x)$ *is true iff* $\sim\varphi(a)$ *is true for every parameter a, or, equivalently, that* $\sim\exists x\varphi(x)$ *and* $\forall x\sim\varphi(x)$ *are logically equivalent (i.e. always have the same truth value);*

(2) $\sim\forall x\varphi(x)$ *is true iff* $\sim\varphi(a)$ *is true for at least one parameter a, or, equivalently, that* $\sim\forall x\varphi(x)$ *and* $\exists x\sim\varphi(x)$ *are logically equivalent.*

Now, let us again recall the uniform notation of *The Beginner's Guide*, in which we let γ be any formula of the form $\forall x\varphi(x)$ or $\sim\exists x\varphi(x)$, and by $\gamma(a)$ we meant $\varphi(a)$ or $\sim\varphi(a)$ respectively. Similarly, we let δ be any formula of the form $\exists x\varphi(x)$ or $\sim\forall x\varphi(x)$, and by $\delta(a)$ we respectively mean $\varphi(a)$, $\sim\varphi(a)$. Thus, in uniform notation, under any first-order valuation v:

F_1': The sentence γ is true under v if and only if for every parameter a, the sentence $\gamma(a)$ is true under v.

F_2': The sentence δ is true under v if and only if the sentence $\delta(a)$ is true under v for at least one parameter a.

The following proposition will be helpful:

Proposition 1. *Suppose that v is a Boolean valuation such that the following two conditions hold (for all sentences γ and δ):*
(1) *if γ is true under v, then for every parameter a, the sentence $\gamma(a)$ is true under v, and*
(2) *if δ is true under v, then $\delta(a)$ is true under v for at least one parameter a.*

Then v is a first-order valuation.

Problem 1. Prove Proposition 1.

We say that a set S of sentences of first-order logic is a *Boolean truth set* if there is a Boolean valuation v such that S is the set of all sentences that are true under v. This is equivalent to saying that for all sentences X and Y, (i) $(\sim X) \in S$ iff $X \notin S$; (ii) $\alpha \in S$ iff $\alpha_1 \in S$ and $\alpha_2 \in S$; (iii) $\beta \in S$ iff $\beta_1 \in S$ or $\beta_2 \in S$.

We call a set of sentences of first-order logic S a *first-order truth set* if and only if it is a Boolean truth set and also, for all γ and δ:
 (i) $\gamma \in S$ iff $\gamma(a) \in S$ for every parameter a; and
 (ii) $\delta \in S$ iff $\delta(a) \in S$ for at least one parameter a.

Alternatively, we could say that a set of sentences S is a *first-order truth set* if and only if there is a first-order valuation v such that S is the set of all sentences that are true under v.

Accordingly, Proposition 1 could be restated as follows:

Proposition 1′. *Suppose S is a* Boolean truth set *satisfying the following two conditions (for all sentences γ and δ):*
(1) *if $\gamma \in S$, then for every parameter a, the sentence $\gamma(a) \in S$, and*
(2) *if $\delta \in S$, then $\delta(a) \in S$ for at least one parameter a.*
Then S is a first-order truth set.

Recall that we say that a (first-order) valuation v *satisfies* a set of sentences S if all elements of S are true under v. Also, a *pure sentence* is a sentence without parameters.

Now for the definition of a magic set. By a *magic set M* we mean a set of sentences that satisfies the following two conditions:
M_1: Every Boolean valuation that satisfies M is also a first-order valuation.
M_2: For every finite set S_0 of *pure* sentences, and every finite subset M_0 of M, if S_0 is first-order satisfiable, so is $S_0 \cup M_0$.

We shall soon prove that magic sets exist (which the reader might find surprising!). But first you will see that one of the important features of magic sets emerges from the following considerations:

We say that a sentence X is *tautologically implied* by a set of sentences S when X is true under all Boolean valuations that satisfy S. We shall call S a *truth-functional basis* for first-order logic if for any *pure* sentence X, the sentence X is valid if and only if X is tautologically implied by some finite subset S_0 of S (which is equivalent to saying that there are finitely many elements $X_1, \ldots, X_n$ of S such that the sentence $(X_1 \wedge \ldots \wedge X_n) \supset X$ is a tautology).

Note: "Truth-functionally imply" and "tautologically imply" have the same meaning.

Theorem 1. *Every magic set is a truth-functional basis for first-order logic.*

Problem 2. Prove Theorem 1. (**Hint**: Use a fact proved in *The Beginner's Guide* [Smullyan, 2014], Chapter 6, Problem 8, namely that if X is tautologically implied by S, then it is tautologically implied by some finite subset of S. This fact is an easy corollary of the compactness theorem for propositional logic, namely that if every finite subset of a denumerable set S is satisfied by some Boolean valuation, then S is satisfied by some Boolean valuation.)

Now we turn to the proof that magic sets exist. In Chapter 9 of *The Beginner's Guide* [Smullyan, 2014] we defined a finite set of sentences to be *regular* if its elements are each of the form $\gamma \supset \gamma(a)$ or $\delta \supset \delta(a)$, and the elements can be placed in a sequence such that for each term of the sequence, if it is of the form $\delta \supset \delta(a)$, then a does not occur in δ or in any earlier term of the sequence. We showed that for any pure sentence X, if X is valid, then there exists a regular set $\{X_1, \ldots, X_n\}$ such that $(X_1 \wedge \ldots \wedge X_n) \supset X$ is a tautology. We also showed that for any regular set R and any set S of pure sentences, if S is (first-order) satisfiable, so is $R \cup S$.

We now define a *denumerably infinite* set S to be regular if every finite subset of S is regular. An infinite regular set M will be called *complete* if for every γ, the sentence $\gamma \supset \gamma(a)$ is in M for every parameter a, and if for every δ, the sentence $\delta \supset \delta(a)$ is in M for at least one parameter a.

Theorem 2. *Every complete regular set is a magic set.*

Problem 3. Prove Theorem 2. [**Hint**: Proposition 1 will now be helpful.]

Finally, we show that there exists a complete regular set (which completes the proof that there exists a magic set): We enumerate all δ-sentences in some denumerable sequence $\delta_1, \delta_2, \ldots, \delta_n, \ldots$. We then consider all our parameters in some enumeration $b_1, b_2, \ldots, b_n, \ldots$. We now let a_1 be the first parameter in the enumeration of parameters that does not occur in δ_1, we let a_2 be the first parameter of the enumeration of parameters that occurs neither in δ_1 nor in δ_2 and so forth, i.e. for each n we let a_n be the first parameter of the enumeration of parameters that occurs in none of the terms $\delta_1, \delta_2, \ldots, \delta_n$. We let R_1 be the set of all sentences $\delta_n \supset \delta_n(a_n)$. We let R_2 be the set of all $\gamma \supset \gamma(a)$, for all γ and every a. Then $R_1 \cup R_2$ is a complete regular set.

Applications of Magic Sets

The mere existence of magic sets (even those not necessarily regular) yields particularly simple and elegant proofs of the Skolem–Löwenheim Theorem and of the first-order compactness theorem as a consequence of the compactness theorem for propositional logic. We will see these proofs in a moment.

Note: But first a reminder of something from *The Beginner's Guide* [Smullyan, 2014]. It was noted there that if we consider the set of all true sentences under a first-order valuation v, and look at the universe in which the individuals to which predicates apply as consisting of all the parameters, we can define an *interpretation* in which each n-ary predicate is true precisely on the n-tuples of parameters on which the valuation v says it is true (and false on the rest of n-tuples of parameters). This is called a (first-order) *interpretation* in the (denumerable) domain of the parameters, and because of this we can say a set of sentences S is first-order satisfiable *in the domain of the parameters* whenever there is a first-order valuation that satisfies S. Although we are not using the concept of an interpretation of first-order logic at the moment here, it is easy to see that any interpretation of first-order logic simultaneously defines a first-order valuation over the denumerable domain of the parameters.

Now suppose S is a denumerable set of pure sentences (no free variables, no parameters) such that every finite subset of S is first-order satisfiable. Here is an alternative proof that S is first-order satisfiable in the denumerable domain of the parameters (which simultaneously yields the compactness theorem for first-order logic, and the Skolem–Löwenheim Theorem, since if S is first-order satisfiable, then obviously so is every finite subset of S). The proof we now give uses the compactness theorem for propositional logic.

Let S be as above and let M be any magic set. Let K be any finite subset of $S \cup M$. We first show that K is first-order satisfiable. Well, K is the union $S_0 \cup M_0$ of some finite subset S_0 of S with some finite subset M_0 of M. Since S_0 is first-order satisfiable, so is $S_0 \cup M_0$ [by property M_2 of a magic set]. Since $S_0 \cup M_0$ is first-order satisfiable, by definition there is a first-order valuation v that satisfies $S_0 \cup M_0$, and v is obviously also a Boolean valuation [since every first-order valuation is]. This proves that every finite subset K of $S \cup M$ is truth-functionally satisfiable. Hence, by the compactness theorem for *propositional* logic, the entire set $S \cup M$ is truth-functionally satisfiable. Thus there is a Boolean valuation v that satisfies $S \cup M$, and since v satisfies M, it must be a first-order valuation [by property M_1 of a magic set]. Thus, by what we said above about the relationship between the first-order interpretations of *The Beginner's Guide* [Smullyan, 2014] and first-order valuations, all sentences of S are true (simultaneously satisfiable) in an interpretation over the denumerable domain of the parameters.

We recall the Regularity Theorem, which is that for every valid pure sentence X, there is a (finite) regular set R which *truth-functionally* implies X. (This is a somewhat simplified version of what we proved in *The Beginner's Guide*. We will return to the full version of the Regularity Theorem later in this chapter.) We proved this using the completeness theorem for tableaux, namely we showed how, from a closed tableau for $\sim X$, such a regular set could be found (we took R to be the set of all $\gamma \supset \gamma(a)$ such that $\gamma(a)$ was inferred from γ by Rule C, together with all sentences $\delta \supset \delta(a)$ such that $\delta(a)$ was inferred from δ by Rule D). A much simpler proof of the Regularity Theorem is easily obtained from the existence of a *complete* regular set (which is also a magic set).

Problem 4. How?

II.   Gentzen Sequents and Some Variants

In Part III of this chapter, we will state and prove an important result known as Craig's Interpolation Lemma, which has significant applications that we will also consider. Craig's original proof of this result was quite complicated, and many simpler proofs of this have subsequently been given. One of these which I will give uses a variant of some axiom systems due to Gerhard Gentzen [1934, 1935], to which we now turn.

Background

We first need to speak of *subformulas*. In propositional logic, the notion of *immediate subformula* is given explicitly by the following conditions:

I_0: Propositional variables have no immediate subformulas.

I_1: $\sim X$ has X as its only immediate subformula.

I_2: $X \wedge Y$, $X \vee Y$, $X \supset Y$ have X and Y as immediate subformulas, and no other immediate subformulas.

For first-order logic, the notion of *immediate subformula* is given explicitly by the following conditions (we recall that an *atomic formula* of first-order logic is a formula which contains no logical connectives or quantifiers):

I_0': Atomic formulas have no immediate subformulas.

I_1': Same as I_1.

I_2': Same as I_2.

I_3': For any parameter a and variable x, $\varphi(a)$ is an immediate subformula of $\forall x \varphi(x)$ and of $\exists x \varphi(x)$.

Both in propositional logic and first-order logic, the notion of *subformula* is implicitly defined by the following conditions:

S_1: If Y is an immediate subformula of X, or is identical to X, then Y is a subformula of X.

S_2: If Y is an immediate subformula of X, and Z is a subformula of Y, then Z is a subformula of X.

The above implicit definition can be made explicit as follows: Y is a subformula of X if and only if either $Y = X$ or there is a finite sequence of formulas whose first term is X and whose last term is Y and is such that each term of the sequence other than the first is an immediate subformula of the preceding term.

Now, G. Gentzen was out to find *finitary proofs* of the consistency of various systems (finitary in the sense that the proofs did not involve the use of infinite sets). For this purpose, he needed a proof procedure that required that any proof of a formula X used only subformulas of X. Such a proof procedure is said to obey the *subformula principle*. Of course, tableaux using signed formulas obey the subformula principle (and these are closely related to the Gentzen systems, as we will see). Tableaux using unsigned formulas, almost, but not quite, obey this principle — they use either subformulas or negations of them. [For example $\sim X$ is not necessarily a subformula of $\sim(X \vee Y)$, but is the negation of one.]

Gentzen took a new symbol "$\rightarrow$" and defined a *sequent* to be an expression of the form $\theta \rightarrow \Gamma$, where each of θ, Γ is a finite, possibly empty, sequence of formulas. Such an expression is interpreted to mean that if all terms of θ are true, then at least one term of Γ is true. That is, under any valuation v, the sequent $\theta \rightarrow \Gamma$ is true under v iff it is the case that if all terms of θ are true, then at least one term of Γ is true. A sequent is called a *tautology* if it is true under all Boolean valuations, and *valid* if it is true under all first-order valuations. If θ is a non-empty sequence $X_1, \ldots, X_n$ and Γ is a non-empty sequence $Y_1, \ldots, Y_k$, then the sequent $X_1, \ldots, X_n \rightarrow Y_1, \ldots, Y_k$ is logically equivalent to the formula $(X_1 \wedge \ldots \wedge X_n) \supset (Y_1 \vee \ldots \vee Y_k)$. If θ is the empty sequence, then $\theta \rightarrow \Gamma$ is written as $\emptyset \rightarrow \Gamma$, where $\emptyset$ is the empty set, or more simply $\rightarrow \Gamma$.

Now, if Γ is a non-empty sequence, then the sequent $\rightarrow \Gamma$ (which is $\emptyset \rightarrow \Gamma$) is read "if all elements of the empty set are true, then at least one element of Γ is true." Well, all elements of the empty set *are* true (since there are no elements of the empty set), and so the sequent $\rightarrow \Gamma$ is logically equivalent to the proposition that at least one element of Γ is true. Thus a sequent $\rightarrow Y_1, \ldots, Y_k$ is logically equivalent to the formula $Y_1 \vee \ldots \vee Y_k$.

Now, suppose that the right side Γ of the sequent $\theta \rightarrow \Gamma$ is empty. The sequent is then more simply written as $\theta \rightarrow$. It is interpreted to mean that if all the terms of the sequence θ are true, then at least one element of the empty set $\emptyset$ is true. Well, it is obviously *not* the case that some element of the empty set is true, so that $\theta \rightarrow$ expresses the proposition that it is not the case that all terms of the sequence θ are true. If θ is a non-empty sequence $X_1, \ldots, X_n$, then the sequent $X_1, \ldots, X_n \rightarrow$ is logically equivalent to the formula $\sim(X_1 \wedge \ldots \wedge X_n)$.

Finally, the $\rightarrow$ written alone is a sequent, and abbreviates $\emptyset \rightarrow \emptyset$, where $\emptyset$ is the empty set. What does $\rightarrow$ by itself mean? It means that if all elements of the empty set are true, then at least one element of the empty set is true. Well, all elements of the empty set are vacuously true, but is it obvious that it is not the case that at least one element set if true. Thus the sequent $\rightarrow$ is simply false (under all valuations).

At this point, I must tell you an amusing incident: I was once giving a lecture on Gentzen sequents to a mathematics group. I started with Gentzen sequents in which both sides of the arrow were non-empty, and after having erased, first the left side, and then the right side, I was left with just the arrow alone on the board. After explaining that it meant falsehood, the logician William Craig (the inventor of Craig's Interpolation Lemma) was in the audience, and with his typical sense of humor, raised his hand and asked, "And what does it mean if you write nothing on the board?"

In more modern Gentzen type systems, the sequents are of the form $U \rightarrow V$, where U and V are finite *sets* of formulas, instead of sequences of formulas, and this is the course we will take.

Gentzen Type Axiom Systems

We shall first consider a Gentzen type axiom system G_0 for *propositional logic*. For any finite set S of formulas and any formulas $X_1, \ldots, X_n$, we abbreviate $S \cup \{X_1, \ldots, X_n\}$ by $S, X_1, \ldots, X_n$. Thus,

$$U, X_1, \ldots, X_n \rightarrow V, Y_1, \ldots, Y_k$$

abbreviates

$$U \cup \{X_1, \ldots, X_n\} \rightarrow V \cup \{Y_1, \ldots, Y_k\}.$$

The system G_0 for propositional logic has only one axiom scheme and the eight inference rules below. Here X and Y are formulas of propositional logic, and U and V are sets of formulas of propositional logic.

Axiom Scheme of the system G_0

$$U, X \rightarrow V, X$$

Rules of the system G_0

Conjunction

$$C_1 : \frac{U, X, Y \to V}{U, X \wedge Y \to V}$$

$$C_2 : \frac{U \to V, X \qquad U \to V, Y}{U \to V, X \wedge Y}$$

Disjunction

$$D_1 : \frac{U \to V, X, Y}{U \to V, X \vee Y}$$

$$D_2 : \frac{U, X \to V \qquad U, Y \to V}{U, X \vee Y \to V}$$

Implication

$$I_1 : \frac{U, X \to V, Y}{U \to V, X \supset Y}$$

$$I_2 : \frac{U \to V, X \qquad U, Y \to V}{U, X \supset Y \to V}$$

Negation

$$N_1 : \frac{U, X \to V}{U \to V, \sim X}$$

$$N_2 : \frac{U \to V, X}{U, \sim X \to V}$$

The inference rules introduce the logical connectives. There are two rules for each connective — one to introduce the connective on the left side of the arrow, the other to introduce the connective on the right side of the arrow. As example of how the rules are applied, the conjunction rule C_1 is that from the sequent $U, X, Y \to V$ one can infer the sequent $U, X \wedge Y \to V$. As for Rule C_2, it says that from the two sequents $U \to V, X$ and $U \to V, Y$ one can infer the sequent $U \to V, X \wedge Y$.

It should be obvious that the axioms of G_0 are tautologies, and we leave it to the reader to verify that the inference rules preserve tautologies, in other words, that in any application of an inference rule, the conclusion is truth-functionally implied by the premises. It therefore follows (by mathematical induction) that we have: All provable sequents of G_0 are tautologies; thus the system G_0 is correct.

We will soon prove that the system G_0 is complete, i.e. that all tautologies are provable in G_0.

Problem 5. Show that the sequent $U, X \to V$ is logically equivalent to $U \to V, \sim X$ and that the sequent $U \to V, X$ is logically equivalent to $U, \sim X \to V$.

The System G_0 in Uniform Notation

Let S be a non-empty set $\{TX_1, \ldots, TX_n, FY_1, \ldots, FY_k\}$ of signed formulas. By $|S|$ we shall mean the sequent $X_1, \ldots, X_n \to Y_1, \ldots, Y_k$. We call $|S|$ the *corresponding sequent* of S. We leave it to the reader to verify the important fact that S is unsatisfiable if and only if its corresponding sequent $|S|$ is a *tautology* (and, for later use when we get to Gentzen systems for first-order logic), that if the elements of S are signed formulas of first-order logic, then S is unsatisfiable iff $|S|$ is *valid*.

Of course, if S is a set $\{FY_1, \ldots, FY_k\}$ in which there are no formulas signed T, we take $|S|$ to be the sequent $\to Y_1, \ldots, Y_k$, and if S is a set $\{TX_1, \ldots, TX_n\}$ in which there are no formulas signed F, we take $|S|$ to be the sequent $X_1, \ldots, X_n \to$.

More generally, for any set of signed formulas, if U is the set of formulas X such that $TX \in S$, and V is the set of formulas Y such that $FY \in S$, then $|S|$ is the sequent $U \to V$.

The system G_0 in uniform notation is the following:

Axioms

(where S is a set of signed formulas)

$$|S, TX, FX|$$

Inference Rules

$$A : \frac{|S, \alpha_1, \alpha_2|}{|S, \alpha|} \quad (\alpha, \alpha_1, \text{ and } \alpha_2 \text{ are signed formulas})$$

$$B : \frac{|S, \beta_1| \qquad |S, \beta_2|}{|S, \beta|} \quad (\beta, \beta_1, \text{ and } \beta_2 \text{ are signed formulas})$$

Problem 6. Verify that the above uniform system really is the system G_0.

Proofs in Gentzen type systems are usually in tree form, unlike the linear form of proofs in the axiom systems previously considered. Only now, unlike the trees of tableaux, the tree must grow *upwards* if the user wishes to prove a statement by analyzing it from the outside in (as must take place by the tableau rules): the origin is at the bottom and the end points are at the top. The origin is the sequent to be proved, and the end points are the axioms used in the proof. As an example, here is a proof in tree form of the sequent $p \supset q \rightarrow \sim q \supset \sim p$.

> (1) $p \rightarrow q, p$ (2) $p, q \rightarrow q$
>
> (3) $p, p \supset q \rightarrow q$ [from (1) and (2), by Rule I_2]
>
> (4) $p \supset q \rightarrow q, \sim p$ [from (3), by Rule N_1]
>
> (5) $p \supset q, \sim q \rightarrow \sim p$ [from (4), by Rule N_2]
>
> (6) $p \supset q \rightarrow \sim q \supset \sim p$ [from (5), by Rule I_1].

Completeness of G_0

We now wish to show that all tautologies are provable in the Gentzen system G_0 for propositional logic.

By a *tableau proof* of a sequent $X_1, \ldots, X_n \rightarrow Y_1, \ldots, Y_k$ is meant a closed tableau proof showing that the set $\{TX_1, \ldots, TX_n, FY_1, \ldots, FY_k\}$ is unsatisfiable (if the tableau is started with all these formulas, the tableau can be extended by the tableau rules for propositional logic in such a way that every branch closes). We will now show how, from a tableau proof of a sequent, we can obtain a proof of the sequent in the Gentzen system G_0. To this end, it will be useful to first introduce, as an intermediary, another type of tableau that I will now call a *block tableau* (this was called a *modified block tableau* in my book *First-Order Logic* [Smullyan, 1968, 1995]). The *analytic tableaux*, which we have studied so far, are variants of the tableaux of Evert Beth [1959], whereas the block tableaux, to which we now turn, are variants of the tableaux of J. Hintikka [1955]. Like the tableaux of Hintikka, the points of the tree in a block tableau proof (this time again with the origin at the top, as in analytic tableaux) are not single formulas, but are finite sets of formulas, which we call the *blocks* of the tableaux. And what we do at any stage of the construction depends solely on the end points of the tree (rather than on the branch as a whole, as in analytic tableaux).

We will first consider block tableaux for propositional logic. By a block tableau for a finite set K of signed formulas of propositional logic, we mean a tree constructed by placing the set K at the origin, and then continuing according to the following rules (here, for a set S and formulas X, Y, we abbreviate the sets $S \cup \{X\}$ and $S \cup \{X\} \cup \{Y\}$ by $\{S, X\}$ or $\{S, X, Y\}$ respectively):

Block Tableau Inference Rules for Propositional Logic

A':
$$\{S, \alpha\}$$
$$|$$
$$\{S, \alpha_1, \alpha_2\}$$

B':
$$\{S, \beta\}$$

$$\{S, \beta_1\} \quad \{S, \beta_2\}$$

In words, these rules are:

A': To any end point $\{S, \alpha\}$ on the tree, we subjoin $\{S, \alpha_1, \alpha_2\}$ as sole successor.

B': To any end point $\{S, \beta\}$ on the tree, we simultaneously subjoin $\{S, \beta_1\}$ as left successor and $\{S, \beta_2\}$ as right successor.

We call a block tableau *closed* if each end point contains some element and its conjugate.

Changing from an analytic tableau to a block tableau is a relatively simple matter. To begin with, in constructing an analytic tableau, when using Rule A, and inferring α_1 from α, subjoin α_2 immediately after it. That is, use Rule A in the form:

$$\alpha$$
$$|$$
$$\alpha_1$$
$$\alpha_2$$

Now, to start a block tableau for a sequent $X_1, \ldots, X_n \to Y_1, \ldots, Y_k$, put the set $\{TX_1, \ldots, TX_n, FY_1, \ldots, FY_k\}$ at the origin, for we wish to prove that set of formulas is unsatisfiable. This concludes stage 0 of the construction of the block tableau corresponding to the analytic tableau that begins as follows:

$$TX_1$$

.

.

.

$$TX_n$$
$$FY_1$$

.

.

.

$$FY_k$$

Then, for each n, after completion of stage n for the two tableaux, when we use Rule A (Rule B) in the analytic tableau, we use Rule A' (Rule B', respectively) in the block tableau, and that concludes stage $n + 1$ of the constructions. We continue this until both tableaux close (which must happen at some stage, if the set of formulas $\{TX_1, \ldots, TX_n, FY_1, \ldots, FY_k\}$ is unsatisfiable, i.e. the sequent is a tautology). Since we know that the tableau method for propositional logic is complete, this construction shows that the block tableau method for propositional logic is also complete.

As an example, here is a closed analytic tableau for the sequent $p \supset q, q \supset r \rightarrow p \supset r$:

$$T\,p \supset q$$
$$T\,q \supset r$$
$$F\,p \supset r$$
$$T\,p$$
$$F\,r$$

$$\diagup\ \diagdown$$

$$\underline{F\,p}\quad T\,q$$

$$\diagup\ \diagdown$$

$$\underline{F\,q}\quad \underline{T\,r}$$

Here is the corresponding block tableau:

$$\{Tp \supset q,\ Tq \supset r,\ Fp \supset r\}$$

$$\mid$$

$$\{Tp \supset q,\ Tq \supset r,\ Tp, Fr\}$$

$$\underline{\{Fp, Tq, \supset r, Tp, Fr\}} \qquad \{Tq, Tq, \supset r, Tp, Fr\}$$

$$\underline{\{Tq, Fq, Tp, Fr\}} \qquad \underline{\{Tq, Tr, Tp, Fr\}}$$

To convert a block tableau to a proof in the Gentzen system G_0, just replace each block

$$\{TX_1, \ldots, TX_n,\ FY_1, \ldots, FY_k\}$$

by the sequent $X_1, \ldots, X_n \to Y_1, \ldots, Y_k$, and the resulting tree, *when turned upside down*, is a proof in the Gentzen system.

For example, from the above closed block tableau, we get the following proof in the Gentzen system:

$$q, p \to q, r \qquad q, r, p \to r$$

$$q \supset r, p \to p, r \qquad q, q \supset r, p \to r$$

$$p \supset q, q \supset r, p \to r$$

$$\mid$$

$$p \supset q, q \supset r \to p \supset r$$

Remarks. It is not surprising that this conversion of a block tableau to a proof in the Gentzen system works, because the block tableau rules are essentially the rules of the Gentzen system G_0 turned upside down. That is, in Rule A' of block tableaux

$$\{\, S, \alpha \,\}$$

$$|$$

$$\{\, S, \alpha_1, \alpha_2 \,\}$$

if we replace the premise and conclusion by the corresponding sequents, we get the rule

$$|\, S, \alpha \,|$$

$$|$$

$$|\, S, \alpha_1, \alpha_2 \,|$$

which is Rule A of the system G_0 *turned upside down.* Likewise, if in Rule B' of block tableaux

$$\{\, S, \beta \,\}$$

$$\diagup \diagdown$$

$$\{\, S, \beta_1 \} \qquad \{\, S, \beta_2 \}$$

we replace the premise and conclusions by the corresponding sequents, we get the rule

$$|\, S, \beta \,|$$

$$\diagup \diagdown$$

$$|\, S, \beta_1 | \qquad |\, S, \beta_2 |$$

which is Rule B of the system G_0 turned upside down.

A Gentzen Type System G_1 for First-Order Logic

For the Gentzen system G_1, we take the axioms and inference rules of the system G_0 (only now applied to formulas of first-order logic instead of formulas of propositional logic), and add the following inference rules:

$$\forall_1: \quad \frac{U, \varphi(a) \rightarrow V}{U, \forall\, x\varphi(X) \rightarrow V} \qquad \forall_2: \quad \frac{U \rightarrow V, \varphi(a)}{U \rightarrow V, \forall\, x\varphi(X)^*}$$

$$\text{*Providing } a \text{ does not occur}$$
$$\text{in the conclusion sequent.}$$

$$\exists_1: \quad \dfrac{U \to V, \varphi(a)}{U \to V, \exists\, x\varphi(x)} \qquad \exists_2: \quad \dfrac{U, \varphi(a) \to V}{U, \exists\, x\varphi(x) \to V^*}$$

*Providing a does not occur
in the conclusion sequent.

One proves the completeness of G_1 from the completeness of first-order block tableau in essentially the same manner as in propositional logic. One first makes a block tableau, whose rules are those of G_0 (with sequents replaced by sets of signed formulas and turned upside down) together with:

$$C: \quad \{S, \gamma\}$$

$$|$$

$$\{S, \gamma, \gamma(a)\}$$

$$D: \quad \{S, \delta\}^*$$

$$|$$

$$\{S, \delta, \delta(a)\}$$

*Providing a does not occur
in the antecedent sequent.

One then converts a closed block tableau into a proof in the Gentzen system G_1 as before, i.e. one replaces each block $\{S\}$ by the sequent $|S|$ and turns the resulting tree upside down.

The System GG

In some of the inference rules of the systems G_0 and G_1 (specifically the negation and implication rules), one transfers a formula from one side of the arrow to the other (or at least incorporates it into some formula on the other side). We need a modification of the system in which this does not happen. This modification, which we will call GG, has a certain symmetry with respect to the two sides of the arrow, and was called a *symmetric Gentzen system* in Smullyan [1968, 1995].

Here is the system GG, which is admirably suited to uniform notation (since we are dealing with sequents here, all formulas refer to *unsigned sentences*):

Axiom Schemes of GG

$$U, X \to V, X$$
$$U, X, {\sim}X \to V$$
$$U \to V, X, {\sim}X$$

Rules of GG

(A)
$$\frac{U, \alpha_1, \alpha_2 \to V}{U, \alpha \to V} \qquad \frac{U \to V, \beta_1, \beta_2}{U \to V, \beta}$$

(B)
$$\frac{U, \beta_1 \to V \quad U, \beta_2 \to V}{U, \beta \to V} \qquad \frac{U \to V, \alpha_1 \quad U \to V, \alpha_2}{U \to V, \alpha}$$

(C)
$$\frac{U, \gamma(a) \to V}{U, \gamma \to V} \qquad \frac{U \to V, \delta(a)}{U \to V, \delta}$$

(D)
$$\frac{U, \delta(a) \to V}{U, \delta \to V^*} \qquad \frac{U \to V, \gamma(a)}{U \to V, \gamma^*}$$

*Providing a does not occur
in the conclusion sequent.

The Completeness of GG

We need another variant of analytic tableaux to prove the completeness of
GG. In some of the rules of analytic tableaux, a conclusion of a rule has a
different sign than that of the premise (which reflects the fact that in the
corresponding Gentzen system, a sentence might change from one side of
the arrow to the other). We now need an analytic tableau system in which
this does not happen.

By an *altered tableau* we mean one constructed according to the following
rules:

$$\begin{array}{c} T {\sim}{\sim}X \\ | \\ T X \end{array} \qquad\qquad \begin{array}{c} F {\sim}{\sim}X \\ | \\ F X \end{array}$$

$$\begin{array}{c} T X \wedge Y \\ | \\ T X \\ T Y \end{array} \qquad\qquad \begin{array}{c} F X \wedge Y \\ \diagup\diagdown \\ F X \quad F Y \end{array}$$

$$T\, X \vee Y$$

$$\diagup \diagdown$$

$$T\,X \quad T\,Y$$

$$F\,X \vee Y$$
$$|$$
$$F\,X$$
$$F\,Y$$

$$T\,X \supset Y$$

$$\diagup \diagdown$$

$$T \sim X \quad T\,Y$$

$$F\,X \supset Y$$
$$|$$
$$F \sim X$$
$$F\,Y$$

$$T \sim (X \wedge Y)$$

$$\diagup \diagdown$$

$$T \sim X \quad T \sim Y$$

$$F \sim (X \wedge Y)$$
$$|$$
$$F \sim X$$
$$F \sim Y$$

$$T \sim (X \vee Y)$$
$$|$$
$$T \sim X$$
$$T \sim Y$$

$$F \sim (X \vee Y)$$

$$\diagup \diagdown$$

$$F \sim X \quad F \sim Y$$

$$T \sim (X \supset Y)$$
$$|$$
$$T\,X$$
$$T \sim Y$$

$$F \sim (X \supset Y)$$

$$\diagup \diagdown$$

$$F\,X \quad F \sim Y$$

$$T \forall\, x \varphi(x)$$
$$|$$
$$T\, \varphi(a)$$

$$F \exists\, x \varphi(x)$$
$$|$$
$$F\, \varphi(a)$$

$$T \exists\, x \varphi(x)^{*}$$
$$|$$
$$T\, \varphi(a)$$

$$F \forall\, x \varphi(x)^{*}$$
$$|$$
$$F\, \varphi(a)$$

$$T \sim \exists\, x \varphi(x)$$
$$|$$
$$T \sim \varphi(a)$$

$$F \sim \forall\, x \varphi(x)$$
$$|$$
$$F \sim \varphi(a)$$

$$T \sim \exists x \varphi(x) \qquad\qquad F \sim \forall x \varphi(x)$$
$$| \qquad\qquad\qquad\qquad |$$
$$T \sim \varphi(a) \qquad\qquad\qquad F \sim \varphi(a)$$

$$T \sim \forall x \varphi(x)^* \qquad\qquad F \sim \exists x \varphi(x)^*$$
$$| \qquad\qquad\qquad\qquad |$$
$$T \sim \varphi(a) \qquad\qquad\qquad F \sim \varphi(a)$$

In the quantification rules (above and below), the asterisk "*" means that the parameter in the conclusion must not appear in the premise.

Here are the rules of the system *altered tableaux* in uniform notation (where the α, β, γ and δ are of course unsigned formulas):

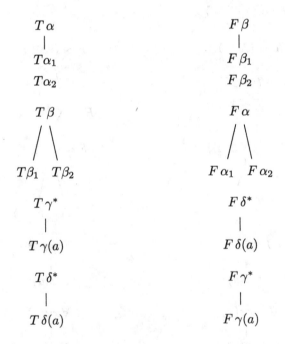

For altered tableaux, a set of signed formulas is called *closed* if it contains either some TX and FX or some TX and $T\sim X$ or some FX and $F\sim X$. A branch is called *closed* if the set of sentences on the branch is a closed set, and an altered tableau is called *closed* if every branch is closed.

We will need to know that the altered tableau method is both correct and complete, i.e. that there is a closed altered tableau for a set S if and only if S is unsatisfiable. To this end, we can save half the labor by the following notation, which might aptly be dubbed a super-unifying notation.

We let A be any signed sentence of the form $T\alpha$ or $F\beta$. If it is the former, we let $A_1 = T\alpha_1$ and let $A_2 = T\alpha_2$; if the latter, we let $A_1 = F\beta_1$ and let $A_2 = F\beta_2$. We let B be any sentence of the form $T\beta$ or $F\alpha$. If it is the former, we let B_1, B_2 be $T\beta_1$ and $T\beta_2$ respectively; if the latter, we let B_1, B_2 be $F\alpha_1$ and $F\alpha_2$, respectively. We let C be any sentence of the form $T\gamma$ or $F\delta$, and we take $C(a)$ to be $T\gamma(a)$, $F\delta(a)$, respectively. We let D be any sentence of the form $T\delta$ or $F\gamma$, and we take $D(a)$ to be $T\delta(a)$, $F\gamma(a)$, respectively. With this notation, the altered tableau rules for first-order logic are now most succinctly expressed as follows:

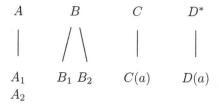

Problem 7. Prove that the altered tableau system is correct and complete, and then finish the proof of the completeness of the Gentzen system GG.

III.   Craig's Lemma and an Application

Craig's Interpolation Lemma

A sentence Z of first-order logic is called an *interpolant* for a sentence $X \supset Y$ if $X \supset Z$ and $Z \supset Y$ are both valid and every predicate and parameter of Z occurs in both X and Y. Craig's celebrated Interpolation Lemma (often simply called Craig's Lemma) is that if $X \supset Y$ is valid, then there is an interpolant for it, provided that Y alone is not valid and X alone is not unsatisfiable [Craig, 1957].

If we allow t's and f's to be part of our formal language, and define "formula" accordingly, then the proviso above is unnecessary, because if $X \supset Y$ is valid and Y alone is valid, then t is an interpolant for $X \supset Y$, or if X alone is unsatisfiable, then f is an interpolant for $X \supset Y$ (as the reader can verify). This is the course we shall take.

Formulas that involve t or f will temporarily be called *non-standard formulas*. As shown in *The Beginner's Guide* [Smullyan, 2014], any non-standard formula is logically equivalent to either a standard formula or to t or to f.

There is a corresponding interpolation lemma for propositional logic. For propositional implication $X \supset Y$, a propositional formula Z is called an interpolant for $X \supset Y$ if $X \supset Z$ and $Z \supset Y$, and if every propositional variable in Z occurs in both X and Y. Craig's Lemma for proposition logic is that if $X \supset Y$ is a tautology, it has an interpolant.

Returning to first-order logic, a formula Z is called an interpolant *for a sequent $U \to V$*, if $U \to Z$ and $Z \to V$ are both valid, and if every predicate and parameter of Z occurs in at least one element of U and at least one element of V. Obviously, Z is an interpolant for a sequent

$$X_1, \ldots, X_n \to Y_1, \ldots, Y_k$$

if and only if Z is an interpolant for the formula

$$(X_1 \land \ldots \land X_n) \supset (Y_1 \lor \ldots \lor Y_k),$$

so that the existence of interpolants for all valid sequents is equivalent to the existence of interpolants for all valid sentences $X \supset Y$. Thus we will consider Craig's Lemma for First-Order Logic in the equivalent form that there exist interpolants for all valid sequents.

We will prove Craig's Lemma for the Gentzen system GG.

Proof of Craig's Interpolation Lemma for the Gentzen System GG

It suffices to show that there is an interpolant for each of the axioms of GG, and that for each of the inference rules of GG, if there is an interpolant for the premise, or interpolants for the two premises, then there in an interpolant for the conclusion.

Exercise. Prove this for the axioms and for Rules A and B (which don't involve the quantifiers).

Proving the claim of the exercise really proves Craig's Interpolation Lemma for propositional logic. Now let's turn to Rules C and D, which do involve the quantifiers.

The proof for Rule D is relatively simple. Suppose that X is an interpolant for $U, \delta(a) \to V$, and that a does not occur in U, δ or V. Then X must also be an interpolant for $U, \delta \to V$ for the following reasons:

Since X is an interpolant of $U, \delta(a) \to V$, then every parameter of X occurs in V (as well as in $U, \delta(a)$). But since a does not occur in V (by hypothesis), then a cannot occur in X. From this, it follows that *all* parameters of X occur in U, δ (as well as in V).

To see this, let b be any parameter that occurs in X. We know that $b \neq a$. We also know that b occurs in V (all parameters of X do). It remains to show that b occurs in U, δ. If b occurs in U, we are done. If not, then b must occur in $\delta(a)$ (since b occurs in U, $\delta(a)$). But the parameters of δ are those of $\delta(a)$ other than a. Since b occurs in $\delta(a)$ and $b \neq a$, then b must occur in δ. This proves that all parameters of X occur in U, δ (as well as in V). Also all predicates of X occur in U, δ and in V since they all occur in $U, \delta(a)$ and in V, and the predicates of δ are the same as those of $\delta(a)$.

Finally, since $U, \delta(a) \to X$ is valid (as well as $X \to V$), and a does not occur in U, δ or in X, then $U, \delta \to X$ is valid (in fact it follows from $U, \delta(a) \to X$ by Rule D). Thus X is an interpolant for the sequent $U, \delta \to V$.

The proof that if X is an interpolant for $U \to V, \gamma(a)$, and a does not occur in U or in $V, \gamma(a)$, then X is an interpolant for $U \to V, \gamma$ is similar to the preceding proof, and is left to the reader. This takes care of Rule D.

The proof for Rule C is much trickier! Suppose that X is an interpolant for $U, \gamma(a) \to V$. Then $U, \gamma(a) \to X$ and $X \to V$ are of course valid, but X may fail to be an interpolant for $U, \gamma \to V$ because the parameter a might occur in X, but not in U, γ. If a does not occur in X, or if a does occur in U, γ, there is no problem. The critical case is one in which a *does* occur in X, but not in U, γ. For this case, we find another interpolant for $U, \gamma \to V$ as follows:

We take some variable x that does not occur in X and we let $\varphi(x)$ be the result of replacing every occurrence of a in X by x. We note that $\varphi(a)$ is X itself. We will show that $\forall x \varphi(x)$ is an interpolant for $U, \gamma \to V$.

Let $\gamma' = \forall x \varphi(x)$. Then $\gamma'(a) = \varphi(a)$, which is X. We already know that $U, \gamma \rightarrow \gamma'(a)$ is valid (because it is $U, \gamma \rightarrow X$), and since a does not occur in U, γ, and obviously not in γ' (which is $\forall x \varphi(x)$), then $U, \gamma \rightarrow \gamma'$ is valid (it is in fact derivable from $U, \gamma \rightarrow \gamma'(a)$ by Rule D of the system GG). Also $\gamma'(a) \rightarrow V$ (which is $X \rightarrow V$) is valid by hypothesis. Hence $\gamma' \rightarrow V$ is valid (by Rule C, taking U to be the empty set). Thus $U, \gamma \rightarrow \gamma'$ and $\gamma' \rightarrow V$ are both valid, and all predicates and parameters of γ' are both in U, γ and in V (since a does not occur in γ'). Thus $\forall x \varphi(x)$, which is γ', is an interpolant for $U, \gamma \rightarrow V$. This completes the proof for the γ rule of Rule C.

As to the δ rule of Rule C, a similar proof, which we leave to the reader, reveals the following: Suppose X is an interpolant for $U \rightarrow V, \delta(a)$. Then if a does not occur in X, or a does occur in V, δ, then X is an interpolant for $U \rightarrow V, \delta$. But if a does occur in X but not in V, δ, then an interpolant for $U \rightarrow V, \delta$ is $\exists x \varphi(x)$, where $\varphi(x)$ is the result of replacing every occurrence of a in X by a variable x that does not occur in X.

This concludes the proof of Craig's Interpolation Lemma.

Beth's Definability Theorem   .

One important application of Craig's Interpolation Lemma is that it provides a neat proof of an important result of the Dutch logician E. Beth [1959]. While discussing Beth's Definability Theorem, I will switch to the language of first-order interpretations that we used frequently in *The Beginner's Guide* [Smullyan, 2014] rather than that of first-order valuations, with which we have spent more time here.

We consider a finite set S of sentences without parameters and involving only predicates $P, P_1, \ldots P_n$ of degree 1. For any sentence X whose predicates are among $P, P_1, \ldots, P_n$, as usual we write $S \vdash X$ to mean that X is a logical consequence of S (i.e. is true in all interpretations that satisfy S). We say that P is *explicitly* definable from $P_1, \ldots, P_n$ with respect to S iff there is a formula $\varphi(x)$ whose predicates are among $P_1, \ldots, P_n$ (and hence do not involve P) such that $S \vdash \forall x (Px \equiv \varphi(X))$. Such a formula $\varphi(x)$ is called an *explicit definition* of P from $P_1, \ldots, P_n$ with respect to S.

There is another type of definability called *implicit definability*, which consists of the following: We take a new predicate P', distinct from any of

$P, P_1, \ldots, P_n$ and which does not occur in any element of S, and we let S' be the result of replacing P by P' in every sentence in S. Then P is said to be *implicitly definable* from $P_1, \ldots, P_n$ with respect to S iff

$$S \cup S' \vdash \forall x(Px \equiv P'x).$$

When this is true, we also say that *the sentences of S implicitly define the predicate P.*

We remark that this condition is equivalent to the condition that for any two interpretations that satisfy S, if they agree on each of $P_1, \ldots, P_n$, then they agree on P. (Two interpretations I_1 and I_2 are said to *agree* on a predicate Q iff they assign the same value to Q.) This equivalence is not necessary for the proof of Beth's theorem, but is of independent interest. A proof of this can be found in my book *Logical Labyrinths* [Smullyan, 2008, 2009, pp. 294–295].

It is relatively easy to show that if P is explicitly definable from $P_1, \ldots, P_n$ with respect to S, then it is implicitly so definable.

Problem 8. Show that if P is explicitly definable from $P_1, \ldots, P_n$ with respect to S, then it is implicitly so definable.

Theorem B. [Beth's Definability Theorem] *If P is implicitly definable from $P_1, \ldots, P_n$ with respect to S, then it is explicitly so definable.*

This theorem is far from obvious, but Craig's Lemma provides a neat and elegant proof of it.

Suppose that P is implicitly definable from $P_1, \ldots, P_n$ with respect to S. Let P' and S' be as previously defined, and assume we have

$$S \cup S' \vdash \forall x(Px \equiv P'x).$$

Let X be the conjunction of the elements of S (the order does not matter), and let X' be the conjunction of the elements of S'. We thus have

$$(X \wedge X') \supset \forall x(Px \equiv P'x).$$

Since the above sentence is valid, then for any parameter a the sentence $(X \wedge X') \supset (Pa \equiv P'a)$ is valid. [We recall that S is a set of sentences without parameters because that is one of the conditions on both explicit and implicit definitions. Thus the sentence $X \wedge X'$ has no parameters.] Hence by propositional logic, the sentence $(X \wedge X') \supset (Pa \supset P'a)$, is valid. Thus, again by propositional logic, so is the following sentence:

$$(X \wedge Pa) \supset (X' \supset P'a). \text{ (Verify this!)}$$

In this sentence, P does not occur in $X' \supset P'a$ and P' does not occur in $X \wedge Pa$. By Craig's Lemma, there is an interpolant Z for the sentence $(X \wedge Pa) \supset (X' \supset P'a)$. All predicates and parameters of Z occur in both $X \wedge Pa$ and in $X' \supset P'a$. Hence neither P nor P' can occur in Z; in fact all the predicates of Z are among $P_1, \ldots, P_n$. Also, Z contains no parameters other than a, since X and X' contain no parameters, as noted above. Let x be a variable that does not occur in Z, and let $\varphi(x)$ be the result of substituting x for all occurrences of a in Z. But $\varphi(a) = Z$, so $\varphi(a)$ is an interpolant for $(X \wedge Pa) \supset (X' \supset P'a)$. Hence the following two sentences are valid:
(1) $(X \wedge Pa) \supset \varphi(a)$.
(2) $\varphi(a) \supset (X' \supset P'a)$.

From (1), we get the valid sentence:
(1') $X \supset (Pa \supset \varphi(a))$.

From (2), we get the valid sentence:
(2') $X' \supset (\varphi(a) \supset P'(a))$.

Since (2') is valid, so is its notational variant:
(2'') $X \supset (\varphi(a) \supset P(a))$.

[If a sentence is valid, it remains valid if we replace any predicate by a new predicate.]

By (1') and (2'') we get the valid sentence:

$$X \supset (Pa \equiv \varphi(a)).$$

Since a does not occur in X, the sentence $X \supset \forall x(Px \equiv \varphi(x))$ is valid. Since X is the conjunction of the elements of S, it then follows that $S \vdash \forall x(Px \equiv \varphi(x))$, and so the formula $\varphi(x)$ explicitly defines P from $P_1, \ldots, P_n$ with respect to S.

This concludes the proof of Beth's Definability Theorem.

Note: The fact that S was assumed finite was not really necessary for the proof. A compactness argument can be used to modify the proof for an infinite set.

IV.   A Unification

Let us review some of the main results that we have proved about first-order logic in *The Beginner's Guide* [Smullyan, 2014] and in this chapter.

T_1: *The Completeness Theorem for Analytic Tableaux.* Every valid formula is provable by the tableau method.

T_2: *The Skolem–Löwenheim Theorem.* For any set S of formulas, if S is satisfiable, then it is satisfiable in a denumerable domain.

T_3: *The Compactness Theorem.* For any infinite set S of formulas, if every finite subset of S is satisfiable, so is S.

T_4: *The Completeness of a First-Order Axiom System.*

T_5: *The Regularity Theorem.* For every valid pure sentence X, there is a (finite) regular set R which *truth-functionally* implies X.

T_6: *Craig's Interpolation Lemma.* If the sentence $X \supset Y$ (of propositional or first-order logic) is valid, there is a sentence Z such that $X \supset Z$ and $Z \supset Y$ are both valid and, for propositional logic, every propositional variable occurring in Z occurs in both X and Y, while, for first-order logic, every predicate and parameter of Z occurs in both X and Y.

In my paper "A Unifying Principle in Quantification Theory" [Smullyan, 1963], I stated and proved a result that yields all of $T_1 - T_6$ as special cases. To this we now turn. What follows is slightly different from the original formulation of 1963.

Consider any property Φ of sentences. A set of sentences that has property Φ will be called Φ-*consistent* (i.e. "consistent" with having the property Φ).

We are now going to be especially interested in certain sets of sentences that have what we will call an *analytic consistency property*, and we will use the Greek symbol Γ to signal that a property of sets is an analytic consistency property. Thus we define a set S of sentences to be Γ-consistent — which means the same as saying that the set S has the analytic consistency property Γ — if the following conditions hold:

Γ_0: S contains no element and its negation (or no element and its conjugate, if we are considering signed formulas).

Γ_1: For any α in S, the sets $S \cup \{\alpha_1\}$ and $S \cup \{\alpha_2\}$ are both Γ-consistent (and hence so is $S \cup \{\alpha_1, \alpha_2\}$).

Γ_2: For any β in S, at least one of the sets $S \cup \{\beta_1\}$ and $S \cup \{\beta_2\}$ is Γ-consistent.

Γ_3: For any γ in S and any parameter a, the set $S \cup \{\gamma(a)\}$ is Γ-consistent.

Γ_4: For any δ in S, the set $S \cup \{\delta(a)\}$ is Γ-consistent for any parameter a that does not occur in any element of S.

Theorem U. [The Unification Theorem] *For any analytic consistency property* Γ, *if* S *is* Γ-*consistent, then* S *is satisfiable in the domain of the parameters.*

One proof of Theorem U is quite similar to the completeness proof for tableaux: Given a Γ-consistent set S, it easily follows that no tableau for S can close (verify this!). Hence a *systematic tableau* for S will contain an open branch that includes all elements of S, and is a Hintikka set, and thus satisfiable in the domain of the parameters.

Discussion. We proved Theorem U using analytic tableaux, although the theorem makes no reference to tableaux. My 1963 proof used neither tableaux nor König's lemma, and is as follows: We shall consider now only the case that S is denumerable (the construction for a *finite* set S is even simpler, and the modification is left to the reader). We shall also assume that only finitely many parameters are found in the formulas of S (if not, we can always introduce a new denumerable group of parameters, and use them along with the parameters in the formulas of S, in the construction ahead).

In what follows, *consistent* will mean Γ-consistent. Given a consistent denumerable set of sentences S, the idea is to define an infinite sequence of formulas whose set of terms is a Hintikka set that contains all the elements of S. The sequence is defined in *stages*: at each stage we have a finite segment $(X_1, \ldots, X_k)$ of our desired infinite sequence, and at the next stage, we extend that sequence to a larger finite sequence $(X_1, \ldots, X_k, \ldots, X_{k+m})$.

First some definitions and notations: Let θ be a finite sequence $(X_1, \ldots, X_k)$. We shall say that θ *is consistent with* S when the set $S \cup \{X_1, \ldots, X_k\}$ is consistent. For any formulas $Y_1, \ldots, Y_m$, by the expression $\theta, Y_1, \ldots, Y_m$ is meant the sequence $(X_1, \ldots, X_k, Y_1, \ldots, Y_m)$.

Given a consistent denumerable set S, let $s_1, \ldots, s_n, \ldots$ be an enumeration of S. We let θ_1 be the unit sequence (s_1). This sequence is obviously consistent with S, since $s_1 \in S$. This concludes the first stage.

Now suppose we have completed the nth stage and have on hand a sequence $(X_1, \ldots, X_k)$, where $k \geq n$. We call this sequence θ_n, and we assume it to be consistent with S. We then extend θ_n to a finite sequence θ_{n+1} in a manner depending on the nature of X_n:

(a) If X_n is an α, we take θ_{n+1} to be θ_n, α_1, α_2, s_{n+1}.

(b) If X_n is an β, then either θ_n, β_1 or θ_n, β_2 is consistent with S. If the former we take θ_{n+1} to be θ_n, β_1, s_{n+1}; if the latter, we take θ_{n+1} to be θ_n, β_2, s_{n+1}.

(c) If X_n is a γ, we let a be the first parameter for which $\gamma(a)$ is not a term of θ_n, and we take θ_{n+1} to be θ_n, $\gamma(a)$, γ, s_{n+1}.

(d) If X_n is a δ, we take θ_{n+1} to be θ_n, $\delta(a)$, s_{n+1}, where a is a parameter new to S and new to θ_n.

This concludes stage $n + 1$ of the sequence.

Since for each n the sequence θ_n is consistent with S, then no formula and its conjugate (or negation, if we are dealing with unsigned formulas) are terms of θ_n. Hence no formula and its conjugate (or negation) are terms of θ. Also it is obvious from the construction that if α is a term of θ, so are α_1 and α_2; and if β is a term of θ, so is either β_1 or β_2; and if γ is a term of θ, then so is $\gamma(a)$ for every parameter a (since we encounter γ denumerably many times); and if δ is a term of θ, then so is $\delta(a)$ for some parameter a. Thus the set of terms of θ is a Hintikka set and includes all elements of S, and so S is denumerably satisfiable in the domain of the parameters.

This construction makes no explicit reference to trees, although θ is in fact the leftmost open branch of a systematic tableau for S. This proof does not use König's Lemma.

Applications: We will provide a sketch of the application of Γ-*consistency* in proofs of the important results listed above. The details will be left to the reader.

T_1: (*Completeness of the tableau method*).

Call a set S *tableau-consistent* if there exists no closed tableau for S. It is easily seen that tableau-consistency is an analytic consistency property. Hence by the Unification Theorem, every tableau-consistent set is satisfiable. Therefore, if a set S is unsatisfiable, there must be a closed tableau for S. Now, if a formula X is valid, then the set $\{\sim X\}$ is unsatisfiable, so that there is a closed tableau for $\sim X$, which means that X is tableau provable.

T_2: (*The Skolem-Löwenheim Theorem*).

Satisfiability itself is easily seen to be an analytic consistency property. Hence by Theorem U, every satisfiable set is satisfiable in the denumerable domain of the parameters.

T_3: (*The Compactness Theorem*).

Call a set S F-consistent if all finite subsets of S are satisfiable. F-consistency is easily seen to be an analytic consistency property. Hence, by Theorem U, if all finite subsets of S are satisfiable, then S is satisfiable.

T_4: (*Completeness of an axiom system for first-order logic*).

Give an axiom system for first-order logic, call a set S *consistent* (with respect to the system) if no element of S is refutable, and no finite set $\{X_1, \ldots, X_n\}$ of elements of S are such that the conjunction $X_1 \wedge \ldots \wedge X_n$ is refutable. For the axiom system given in *The Beginner's Guide* [Smullyan, 2014], or for any other standard system, this consistency is easily shown to be an analytic consistency property. Hence every consistent set is satisfiable and every unsatisfiable set is inconsistent. If now X is valid, then $\sim X$ is unsatisfiable, and hence inconsistent. This means that $\sim X$ is refutable, i.e. $(\sim\sim X)$ is provable, and so is X. Thus every valid formula is provable, and the system is *complete*.

T_5: (*The Regularity Theorem*).

Let us review some definitions and facts: We recall that a (finite) *regular set R* means a set whose members can be arranged in a sequence

$$Q_1 \supset Q_1(a_1), \ldots, Q_n \supset Q_n(a_n),$$

where for each $i \leq n$, Q_i is either a γ or a δ, and if Q_i is a δ, the parameter a_i does not occur in δ, nor in any earlier term of the sequence. Such a parameter (occurring in a $Q_i(a_i)$ where Q_i is a δ in a regular set R) is called a *critical parameter* of R.

We recall that a set S is said to *tautologically* imply a formula X if X is true in all *Boolean* valuations that satisfy S. For S a finite set $\{X_1, \ldots, X_n\}$, this condition is equivalent to the condition that the sentence $(X_1 \wedge \ldots \wedge X_n) \supset X$ is a *tautology* (or, in the case that $n = 1$, that $X_1 \supset X$ is a tautology). A finite set $\{X_1, \ldots, X_n\}$ is said to be *truth-functionally unsatisfiable* if the sentence $\sim(X_1 \wedge \ldots \wedge X_n)$ is a *tautology*.

The Regularity Theorem is that for every valid sentence X, there is a regular set R that *tautologically* implies X, and which is such that no critical parameter of R occurs in X.

For any finite set S (of sentences), an *associate* of S is a regular set R such that $R \cup S$ is *truth-functionally* unsatisfiable, and no critical parameter of R occurs in any element of S. Now the Regularity Theorem is equivalent to the proposition that every unsatisfiable (finite) set S has an associate.

Problem 9. Show that the Regularity Theorem is equivalent to the proposition that every unsatisfiable (finite) set S has an associate.

Let us now call a finite set S *A-consistent* if S has no associate. It can be shown that A-consistency is an analytic consistency property. Hence by Theorem U, for any finite set S, if S has no associate, then S is satisfiable. Hence if S is unsatisfiable, then S has an associate, which proves the Regularity Theorem [by Problem 8].

T_6: (*Craig's Interpolation Lemma*).

By a *partition* $S_1 | S_2$ of a set S we mean a pair of subsets S_1 and S_2 of S such that every element of S is in either S_1 or S_2, but not in both; in other words $S_1 \cup S_2 = S$ and $S_1 \cap S_2 = \emptyset$.

By a *partition interpolant* between S_1 and S_2 we mean a sentence Z such that all predicates and parameters of Z occur in both S_1 and S_2, and $S_1 \cup \{\sim Z\}$ and $S_2 \cup \{Z\}$ are both unsatisfiable. Now, call a set S *Craig-consistent* if there is a partition $S_1 | S_2$ of S such that there is no partition interpolant between S_1 and S_2. Well, Craig-consistency can be shown to be an analytic consistency property. Hence by Theorem U, every Craig-consistent set is satisfiable. Therefore if S is unsatisfiable, then for *every* partition $S_1 | S_2$ of S, there is a partition interpolant between S_1 and S_2.

Now, suppose $X \supset Y$ is valid. Then the set $\{X, \sim Y\}$ is unsatisfiable. Consider the partition $\{X\} | \{\sim Y\}$. Then there is a partition interpolant Z between $\{X\}$ and $\{\sim Y\}$. Thus $\{X, \sim Z\}$ (which is $\{X\} \cup \{\sim Z\}$) and $\{Z, \sim Y\}$ (which is $\{\sim Y\} \cup \{Z\}$) are both unsatisfiable, hence $X \supset Z$ and $Z \supset Y$ are both valid, and of course every predicate and parameter of Z are in both X and Y. Thus Z is an interpolant of $X \supset Y$.

Grand Exercise. Verify that the six properties which I claimed to be analytic consistency properties really are so.

V. A Henkin-Style Completeness Proof

Adolf Lindenbaum (a Polish logician born in 1904 and killed by the Nazis in 1941), proved in the 1920s that every consistent set could be extended to a maximally consistent set, a result that afterwards became known as Lindenbaum's Lemma. Leon Henkin [1949] published a proof of the completeness of various axiom systems for first-order logic by a method which is completely different from the ones we have so far considered, and which incorporates the method Lindenbaum used to prove his lemma.

Henkin's method combines maximal consistency with another condition: A set S of sentences is called *E-complete* (existentially complete) if for any element δ in S, there is at least one parameter a such that $\delta(a) \in S$. Henkin showed that every consistent set S can be extended to a set M that is both maximally consistent and E-complete, and that every such set M is a (first-order) truth set. We will see now that Henkin's method applies to a more general situation.

Recall from Section IV that we are using the phrase "the set S of sentences is Γ-consistent" to mean that S has the property Γ and that Γ is an analytic consistency property. We define a property Δ of sentences to be a *synthetic consistency property*, if it is an analytic consistency property *and* satisfies the additional condition that for any set S having property Δ, and any sentence X, either $S \cup \{X\}$ has the property Δ, or $S \cup \{\sim X\}$ has the property Δ. We will call a set Δ-*consistent* if it has a particular synthetic consistency property Δ. So in what follows "Γ" will be used to designate any analytic consistency property, and "Δ" will be used to designate any synthetic consistency property (but remember that any synthetic consistency property is also an analytic consistency property).

A property P of sets is said to be of *finite character* if for any set S, S has the property P if and only if all finite subsets of S have property P.

Henkin's completeness proof for standard axiom systems for first-order logic goes through for any synthetic consistency property Δ of finite character. In this more general form its statement is the following: for any synthetic consistency property Δ of finite character, any Δ-consistent set S can be extended to a maximally Δ-consistent set M, and this can be done in such a way that the set M will be a first-order truth set.

We will show a suitable modification of Henkin's proof for Δ-consistent properties (synthetic consistency properties) of finite character yields the following Henkin-style proof for Γ-consistent properties of finite character, i.e. for *any analytic consistency property* of finite character).

So, consider any analytic consistency property Γ that is also of finite character. We will show that if S is Γ-consistent, then S can be extended to a set M which is both maximally Γ-consistent and E-complete, and that the set M obtained in this way is a Hintikka set, and hence satisfiable (so that M's subset S is also satisfiable).

Problem 10. Show that if M is both maximally Γ-consistent and E-complete, then M is a Hintikka set.

Note: As previously remarked, for any *synthetic* consistency property Δ of finite character, if M is both maximally Δ-consistent and E-complete, then M is not only a Hintikka set, but even a truth set. Indeed, more generally, any Hintikka set S having the property that for every X, either $X \in S$ or $(\sim X) \in S$, must be a truth set. We leave the proof of this as an exercise for the reader.

Now, given an analytic consistency property Γ of finite character, how do we actually go about extending a given Γ-consistent set S of finite character to a set M which is both maximally Γ-consistent and E-complete? (As in our proof of Theorem U above, we will assume that the number of parameters in S is finite, or that we have an additional denumerable number of parameters available to us, none of which occur in any formula of S, i.e. parameters that we can just add to the parameters in S if necessary.)

Here is Lindenbaum's method of extending a consistent set S to a maximally consistent set M (assuming consistency to be of finite character), which we will extend from the conventional meaning of consistent to our meaning of Γ-consistent (i.e. analytically consistent): We first arrange *all* the sentences of first-order logic in some infinite sequence $X_1, X_2 \ldots, X_n, X_{n+1}, \ldots$. Now consider a Γ-consistent set S. We construct an infinite sequence of sets $S_0, S_1, \ldots, S_n, S_{n+1}, \ldots$ as follows: We take S_0 to be S. Then, for every positive n, assuming S_n already defined, we take S_{n+1} to be S_n if $S_n \cup \{X_n\}$ is *not* consistent; while if $S_n \cup \{X_n\}$ *is* consistent, and if X_n is not some δ, then we take S_{n+1} to be $S_n \cup \{X_n\}$. But if $S_n \cup \{X_n\}$ is consistent and X_n is some δ, then $S_n \cup \{\delta, \delta(a)\}$ is also consistent for any parameter a new to $S_n \cup \{\delta\}$ (since we are dealing with an analytic consistency property). In this case we take S_{n+1} to be $S_n \cup \{\delta, \delta(a)\}$, where a is a parameter new to S_n and to δ. The set M that results from taking the union of all the sets $S_0, S_1, \ldots, S_n, S_{n+1}, \ldots$ is both maximally consistent and E-complete. For it is easily seen to be E-complete by construction (realize that at every stage of the construction, there will always be a new parameter available). And the union M of all the consistent sets S_n is consistent because we are assuming our consistency property to be of finite character. And it is maximally consistent because any formula not already in it was kept out of M because adding it would have made a finite subset of M inconsistent.

Immediately after proving that every consistent set S can be extended to a set M that is both maximally consistent and E-complete, Henkin stated as a corollary the completeness of first-order logic, that is, that every valid formula was provable. But he did use a well-known property of the axiom systems he was considering that does not apply to the slightly different formalization of first-order logic in this volume (his systems included the constants t and f, which we did discuss in the context of propositional logic in *The Beginner's Guide* [Smullyan, 2014], and which can also be employed in first-order logic).

Now, we might try to show the completeness of first-order logic similarly to the way Henkin did by first showing that ordinary consistency is an analytic consistency property (a common definition of an individual sentence X being consistent would be that there exists a formula Z such that $X \supset Z$ is not provable). But far easier for us is to just remind ourselves that in Section IV of this chapter (the Unification section), we showed that both the first-order tableau method of proof and the standard first-order axiomatic systems of proof are complete by defining the appropriate analytic consistency property for each of them.

We now see that there are basically two types of completeness proofs directly applicable to the usual formulations of First-Order Logic. One is along the lines of Lindenbaum and Henkin, which extends a consistent set directly to a truth set, and for this we need the full force of *synthetic consistency properties*. The second type of completeness proof (which is along the lines of Gödel, Herbrand, Gentzen, Beth, and Hintikka) extends a consistent set, not directly to a truth set, but to a Hintikka set, which can be further extended to a truth set. And for the second type of proof, we do not need synthetic consistency, but only analytic consistency.

Solutions to the Problems of Chapter 2

1. We are given that v is a Boolean valuation.

(1) We are given that if γ is true under v, then so is $\gamma(a)$ for every parameter a. In addition, we must prove the converse, i.e. that if $\gamma(a)$ is true under v for every parameter a, then γ is true under v. We shall prove the equivalent proposition that if γ is false under v, then so is $\gamma(a)$ for at least one parameter a.

We first consider the case that γ is of the form $\forall x \varphi(x)$. Now suppose that $\forall x \varphi(x)$ is false under v. Then $\sim\forall x \varphi(x)$ is true under v [since v is a Boolean valuation]. Hence by (2) of the lemma, there is at least one parameter a such that $\sim\varphi(a)$ is true under v. Hence $\varphi(a)$ is false under v, which is what was to be shown [since if γ is $\forall x \varphi(x)$, $\gamma(a)$ is $\varphi(a)$].

Now consider the case that γ is of the form $\sim\exists x \varphi(x)$. Well, suppose that $\sim\exists x \varphi(x)$ is false under v. This means that $\exists x \varphi(x)$ is true under v [since v is a Boolean valuation]. Then by F_2, $\varphi(a)$ is true for at least one parameter a. But then $\sim\varphi(a)$ is false for that parameter. And that is what we needed to prove, since when the γ formula is of the form $\sim\exists x \varphi(x)$, $\gamma(a)$ is $\sim\varphi(a)$.

(2) We are given that if δ is true under v, then so is $\delta(a)$ for at least one parameter a. Again we must prove the converse, i.e. that if $\delta(a)$ is true for at least one parameter a, so is δ. We will prove the equivalent proposition that if δ is false, then so is $\delta(a)$ for every parameter a.

We first consider the case that δ is of the form $\exists x \varphi(x)$. So suppose $\exists x \varphi(x)$ is false. Then $\sim\exists x \varphi(x)$ is true under v [since v is a Boolean valuation]. Then by (1) of the Lemma, $\sim\varphi(a)$ is true for every parameter a. Hence $\varphi(a)$ is false for every parameter a, which was what had to be proved [since $\varphi(a)$ is $\delta(a)$ when δ is of the form suppose $\exists x \varphi(x)$].

Now consider the case that δ is of the form $\sim\forall x \varphi(x)$. Well, suppose $\sim\forall x \varphi(x)$ is false. Then $\forall x \varphi(x)$ is true [since v is a Boolean valuation]. Then by F_1, $\varphi(a)$ is true for every parameter a. This means that, $\sim\varphi(a)$ is false for every parameter a [since v is a Boolean valuation], which is what had to be proved [since when δ is of the form $\sim\forall x \varphi(x)$, $\delta(a)$ is of the form $\sim\varphi(a)$].

2. We must show that every magic set M is a truth-functional basis for first-order logic. Thus we must show that if M is a magic set and X is a pure sentence, then X is valid if and only if X is tautologically implied by some finite subset M_0 of M. We assume M is a magic set.

   (a) Suppose that X is a valid pure sentence. Then X is true under all first-order valuations, and thus also true under all Boolean valuations that satisfy M (because all such valuations are also first-order valuations by condition M_1 in the definition of a magic set). Thus X is tautologically implied by M [by the definition of tautological implication], and hence by some finite subset M_0 of M [by Problem 8 of Chapter 6 of *The Beginner's Guide* [Smullyan, 2014]].

   (b) Conversely, suppose X is a pure sentence that is tautologically implied by some finite subset M_0 of M. Then the set $M_0 \cup \{\sim X\}$ is not truth-functionally satisfiable (i.e. not satisfied by any Boolean valuation), so that $M_0 \cup \{\sim X\}$ is not first-order satisfiable (since every first-order valuation is also a Boolean valuation). If $\sim X$ were first-order satisfiable, then $M_0 \cup \{\sim X\}$ would also be first-order satisfiable [by M_2], but since it isn't, then $\sim X$ is not first-order satisfiable, which means that X is valid.

   This completes the proof.

3. We must show that if M is a *complete regular set* of sentences, then M is a magic set. So let M be a complete regular set of sentences.

   Proof of M_1: We must show that any Boolean valuation that satisfies M is also a first-order valuation. So suppose v is a Boolean valuation that satisfies M. By Proposition 1, to show that v is a first-order valuation, it suffices to show that if γ is true under v, so is $\gamma(a)$ for every parameter a, and that if δ is true under v, then so is $\delta(a)$ for at least one parameter a. Well, suppose γ is true under v. Since M is complete, $\gamma \supset \gamma(a)$ is in M for every parameter a. Thus $\gamma \supset \gamma(a)$ is true under v (since every member of M is true under v). Then $\gamma(a)$ must also be true under v (since v is a Boolean valuation).

   Now suppose δ is true under v. Since M is complete, $\delta \supset \delta(a)$ is in M for *some* parameter a. Thus $\delta \supset \delta(a)$ if true under v (since every member of M is true under v). Then $\delta(a)$ must also be true under v (since v is a Boolean valuation).

Proof of M_2: We must show that for every finite set S of *pure* sentences, and every finite subset M_0 of M, if S is first-order satisfiable, so is $S \cup M_0$.

We first show that for any finite set S of sentences (possibly with parameters), if S is (first-order) satisfiable, then:

(1) $S \cup \{\gamma \supset \gamma(a)\}$ is satisfiable, for any γ and any a;

(2) $S \cup \{\delta \supset \delta(a)\}$ is satisfiable, for any δ and any a that does not occur in δ or in any element of S.

Now (1) is obvious, since $\gamma \supset \gamma(a)$ is valid, hence true under *all* interpretations.

As for (2), let I be an interpretation of all predicates and parameters of S in some non-empty universe U which is such that all elements of S are true under I. Extend I to an interpretation I_1 by assigning values in U to all predicates and parameters of δ which do not occur in S. Now let a be any parameter which occurs neither in δ nor in any element of S. We now wish to extend I_1 to an interpretation I_2 in which $\delta(a)$ has a truth value — that is, we must assign a value k of U to the parameter a and we wish to do so in a way that makes $\delta \supset \delta(a)$ true. Well, if δ is false under I_1, then any k in U will do, since $\delta \supset \delta(a)$ will automatically be true. On the other hand, if δ is true under I_1, then there must be at least one element k in U such that $\delta(a)$ is true when k is assigned to a. Thus, we assign such an element k to a, and then, since this assignment makes $\delta(a)$ true, $\delta \supset \delta(a)$ must also be true.

This proves (1) and (2). Finally, consider a finite subset M_0 of M. Then M_0 is a finite regular set. Arrange M_0 in some regular sequence $Q_1 \supset Q_1(a_1), \ldots, Q_n \supset Q_n(a_n)$, where for each $i \leq n$, Q_i is either a γ or a δ, and if Q_i is a δ, the parameter a_i does not occur in δ, nor in any earlier term of the sequence.

Now let S be any satisfiable finite set of pure sentences. Let

$$S_1 = S \cup \{Q_1 \supset Q_1(a_1)\};$$

And for each $i < n$, let

$$S_{i+1} = S_i \cup \{Q_{i+1} \supset Q_{i+1}(a_{i+1})\}.$$

By (1) and (2) above, S_1 is satisfiable, and for each $i < n$, if S_i is satisfiable, so is S_{i+1}. It then follows by mathematical induction that each of the sets S_1, S_2, ..., S_n is satisfiable. Thus $S \cup M_0$, which is S_n, is satisfiable.

4. Let M be a complete regular set of sentences. By Theorem 2, it is also a magic set, so that by Theorem 1, it is a truth-functional basis for first-order logic. Thus, for any valid pure sentence X there is a finite subset R of M which truth-functionally implies X. The set R is obviously regular.

5. For any set S of formulas, let $Con\,S$ be the conjunction of the elements of S (the order does not matter) if S is non-empty, or t if S is empty (in which case all elements of S are true). In either case, under any interpretation, $Con\,S$ is true if and only if all elements of S are true. Let $Dis\,S$ be the disjunction of all the elements of S if S is non-empty, or f if S is empty. In either case, under any interpretation, $Dis\,S$ is true iff at least one element of S is true. A sequent $U \rightarrow V$ is logically equivalent to the simple formula $Con\,U \supset Dis\,V$.

   We are to show that $U, X \rightarrow V$ is logically equivalent to $U \rightarrow V, {\sim}X$, or what is the same thing, that $Con\,(U, X) \supset Dis\,V$ is equivalent to $Con\,U \supset Dis\,(V, {\sim}X)$. [The proof that $U \rightarrow V, X$ is logically equivalent to $U, {\sim}X \rightarrow V$ is similar.]

   Well, under any interpretation (Boolean valuation), X is either true or false.

   Case 1. Let X be true. Then $Con\,(U, X)$ is logically equivalent to $Con\,(U)$, so that $Con\,(U, X) \supset Dis\,V$ is logically equivalent to $Con\,(U) \supset Dis\,V$. But also, since X is true, ${\sim}X$ is false, so that $Dis\,(V, {\sim}X)$ is logically equivalent to $Dis\,(V)$. Thus $Con\,U \supset Dis(V, {\sim}X)$ is also equivalent to $Con\,U \supset Dis\,(V)$. So in this case, the sequents $U, X \rightarrow V$ and $U \rightarrow V, {\sim}X$ are both equivalent to $U \rightarrow V$, hence equivalent to each other.

   Case 2. Let X be false. Then $Con\,(U, X)$ is false, hence $Con\,(U, X) \supset Dis\,V$ is true. But also, since X is false, then ${\sim}X$ is true, so that $Dis\,(V, {\sim}X)$ is true, and therefore $Con\,U \supset Dis\,(V, {\sim}X)$ is true. Thus in this case the sequents $U, X \rightarrow V$ and $U \rightarrow V, {\sim}X$ are both true, hence equivalent to each other.

6. S is a set TX_1, ..., TX_n, FY_1, ..., FY_k of signed formulas. We let U be the set X_1, ..., X_n and let V be the set $\{Y_1, ..., Y_k\}$.

First, let us consider the axioms $\{S, TX, FX\}$ of the uniform system. The set $\{S, TX, FX\}$ is the set

$$\{TX_1, \ldots, TX_n, FY_1, \ldots, FY_k, TX, FX\}$$

and so $|S, TX, FX|$ is the sequent

$$X_1, \ldots, X_n, X \rightarrow Y_1, \ldots, Y_k, X.$$

This is $U, X \rightarrow V, X$, the axiom of G_0.

As for the inference rules, we must consider the various α and β cases separately. For example, consider the case that α is of the form $FX \supset Y$. Then $\alpha_1 = TX$ and $\alpha_2 = FY$. Then $|S, \alpha_1, \alpha_2|$ is $|TX_1, \ldots, TX_n, FY_1, \ldots, FY_k, TX, FY|$, which is the sequent

$$X_1, \ldots, X_n, X \rightarrow Y_1, \ldots, Y_k, Y,$$

which is $U, X \rightarrow V, Y$, and so for the case $\alpha = FX \supset Y$, Rule A is Rule I_1 of the system G_0.

We leave it to the reader to verify the other cases, specifically that if α is $TX \wedge Y$, $FX \vee Y$, $T{\sim}X$, $F{\sim}X$, then Rule A is respectively Rule C_1, D_1, N_2, N_1 of the system G_0, and if β is $FX \wedge Y$, $TX \vee Y$, $TX \supset Y$, then Rule B is respectively Rule C_2, D_2, I_2.

7. First for the *correctness* of the altered tableau method. We are to show that if there is a closed altered tableau for a set, then the set is really unsatisfiable.

   For any satisfiable finite set S of signed formulas, the following facts hold:

   (1) If $A \in S$ then $S \cup \{A_1\}$ and $S \cup \{A_2\}$ are both satisfiable.
   (2) If $B \in S$ then either $S \cup \{B_1\}$ or $S \cup \{B_2\}$ is satisfiable.
   (3) If $C \in S$ then for every parameter a the set $S \cup \{C(a)\}$ is satisfiable.
   (4) If $D \in S$ then for every parameter a *new to S*, the set $S \cup \{D(a)\}$ is satisfiable.

   These facts can be proved in the same manner as in *The Beginner's Guide* [Smullyan, 2014], replacing $\alpha, \beta, \gamma, \delta$ by A, B, C, D respectively. It then follows that if S is satisfiable, at no stage can an altered tableau for S close. Hence if there is a closed altered tableau for S, then S is unsatisfiable. Thus the altered tableau method is correct.

As for *completeness*, one can make a *systematic* altered tableau for a set in essentially the same manner as for analytic tableaux (in the instructions given in *The Beginner's Guide* [Smullyan, 2014], just replace $\alpha, \beta, \gamma, \delta$ by A, B, C, D respectively). If the systematic altered tableau runs on forever without closing, then it contains at least one infinite open branch, and the set S of the sentences on the branch satisfies the following conditions:

(0) For no X in S is it the case that both TX and FX are both in S, or that TX and $T\sim X$ are both in S or that FX and $F\sim X$ are both in S.

(1) If $A \in S$, so are A_1 and A_2.

(2) If $B \in S$, then $B_1 \in S$ or $B_2 \in S$.

(3) If $C \in S$, then so is $C(a)$ for every parameter a.

(4) If $D \in S$, then so is $D(a)$ for at least one parameter a.

Such a set S is very much like a Hintikka set, and the proof of its satisfiability is very close to that for Hintikka set. About the only significant difference is in the assignment of truth values to the *atomic* sentences. Now, for any *atomic* unsigned sentence X, if $TX \in S$ or $F\sim X \in S$, give X the value *truth* (in which case TX and $F\sim X$ are then true). If either $T\sim X \in S$ or $FX \in S$, then give X the value *false* (in which case $T\sim X$ and FX are then true). [No ambiguity can result, because if $TX \in S$ or $F\sim X \in S$, then neither $T\sim X \in S$ nor $FX \in S$ can hold, and if $T\sim X \in S$ or $FX \in S$, then neither $TX \in S$ nor $F\sim X \in S$ can hold.] If none of TX, $T\sim X$, FX, $F\sim X$ occur in S, then give X the value truth or falsity at will — say truth. Under this valuation, all elements of S can be shown to be true by mathematical induction on degrees, much as in the proof of Hintikka's Lemma. Thus S is satisfiable.

We now see that if the systematic altered tableau for S does not close, then S is satisfiable, hence if S is not satisfiable, then any systematic altered tableau for S must close. This completes the *completeness* proof for the altered tableau method.

To complete the proof of the completeness of the Gentzen system GG, we now need altered *block* tableaux for first-order logic, whose rules in uniform notation are the following:

$$\{S, T\alpha\} \qquad \{S, F\beta\} \qquad \{S, T\beta\} \qquad \{S, F\alpha\}$$

$$\{S, T\alpha_1, T\alpha_2\} \; \{S, F\beta_1, F\beta_2\}$$

$$\{S, T\beta_1\} \quad \{S, T\beta_2\} \quad \{S, F\alpha_1\} \quad \{S, F\alpha_2\}$$

$$\{S, T\gamma\} \qquad \{S, F\delta\} \qquad \{S, T\delta\}^* \qquad \{S, F\delta\}^*$$

$$\{S, T\gamma(a)\} \qquad \{S, F\delta(a)\} \qquad \{S, T\delta(a)\} \qquad \{S, F\gamma(a)\}$$

*Providing a is not in the premise.

We recall that if S is the set $\{TX_1, \ldots, TX_n, FY_1, \ldots, FY_k\}$, by $|S|$ we mean the sequent $X_1, \ldots, X_n \to Y_1, \ldots, Y_k$.

For any block tableau rule of the form

$$S_1$$
$$|$$
$$S_2$$

by the *counterpart* of the rule we mean

$$|S_2|$$
$$\overline{}$$
$$|S_1|$$

(i.e. "from the sequent $|S_2|$, to infer the sequent $|S_1|$"). For any block tableau rule of the form

$$S_1$$
$$/\;\backslash$$
$$S_2 \quad S_3$$

by its counterpart is meant the rule

$$|S_2| \quad |S_3|$$
$$\overline{}$$
$$|S_1|$$

(i.e. "from the sequents $|S_2|$ and $|S_3|$, to infer the sequent $|S_1|$"). Well, the counterparts of the eight block tableau rules are precisely the eight rules of the system GG. To see this, let U be the set of unsigned sentences X such that $TX \in S$ and let V be the set of sentences Y such that $FY \in S$. Consider, for example, the block tableau rule

$$\{S, T\alpha\}$$

$$|$$

$$\{S, T\alpha_1, T\alpha_2\}$$

The counterpart of this rule is

$$\frac{|S, T\alpha_1, T\alpha_2|}{|S, T\alpha|}$$

but this is just

$$\frac{U, \alpha_1, \alpha_2 \to V}{U, \alpha \to V}$$

which is a rule of GG.

Another example: The counterpart of the rule is

$$\{S, F\alpha\}$$

$$\diagup\diagdown$$

$$\{S, F\alpha_1\} \quad \{S, F\alpha_2\}$$

$$\frac{|S, F\alpha_1| \quad |S, F\alpha_2|}{|S, F\alpha|}$$

which is

$$\frac{U \to V, \alpha_1 \quad U \to V, \alpha_2}{U \to V, \alpha}$$

which is a rule of GG.

The reader can verify the cases of the other six rules.

Now if $\mathscr{T}$ is a closed altered block tableau for a set of signed formulas whose counterpart is the sequent $U \to V$, each end point S is closed, and the sequence $|S|$ is an axiom of GG:

    (i) $|S, TX, FX|$ is $U, X \rightarrow V, X$;

    (ii) $|S, TX, T{\sim}X|$ is $U, X, {\sim}X \rightarrow V$;

    (iii) $|S, FX, F{\sim}X|$ is $U \rightarrow V, X, {\sim}X$.

And since the counterparts of the block rules are rules of GG, it follows that if we replace each point S of the tree $\mathscr{T}$ by the sequent $|S|$ and turn the tree upside down, we have a proof of the sequent $U \rightarrow V$ in the system GG. This concludes the proof of the completeness of GG.

8. It is understood that P' is a predicate distinct from each of $P, P_1, \ldots, P_n$, and that P' occurs in no element of S, and that S' is the result of substituting P' for P in every element of S.

    Now suppose that P is explicitly definable from $P_1, \ldots, P_n$ with respect to S. Let $\varphi(x)$ be a formula whose predicates are among $P_1, \ldots, P_n$ but do not involve P and which is such that

$$S \vdash \forall x(Px \equiv \varphi(X)).$$

Then also, of course, $S' \vdash \forall x(P'x \equiv \varphi(X))$. Hence,

$$S \cup S' \vdash \forall x(Px \equiv \varphi(X)) \wedge \forall x(P'x \equiv \varphi(X)),$$

which implies

$$S \cup S' \vdash \forall x((Px \equiv \varphi(X)) \wedge (P'x \equiv \varphi(X))).$$

Therefore $S \cup S' \vdash \forall x(Px \equiv P'x)$ is logically valid, as the reader can verify (say, with a tableau). Thus P' is implicitly definable from $P_1, \ldots, P_n$ with respect to S.

9. (a) Suppose every unsatisfiable set has an associate. Consider any valid sentence X. Then the unit set $\{{\sim}X\}$ is unsatisfiable, hence has an associate R. Since $R \cup \{{\sim}X\}$ is not truth-functionaly satisfiable, then R tautologically implies X [since X must be clearly true in any Boolean valuation satisfying R]. Moreover, X contains no critical parameter of R, which proves the Regularity Theorem.

  (b) Conversely, assume the Regularity Theorem. Let S be a finite set $\{X_1, \ldots, X_n\}$ of sentences which is unsatisfiable. Then the sentence ${\sim}(X_1 \wedge \ldots \wedge X_n)$ is valid. Hence by the assumed Regularity Theorem, there is a regular set R which tautologically implies ${\sim}(X_1 \wedge \ldots \wedge X_n)$ and which is such that no critical parameter of R occurs in any element of ${\sim}(X_1 \wedge \ldots \wedge X_n)$, and thus no critical parameter of R occurs in any element of S. Since R tautologically implies ${\sim}(X_1 \wedge \ldots \wedge X_n)$, then $R \cup \{X_1, \ldots, X_n\}$ is truth-functionally unsatisfiable, i.e. $R \cup S$ is truth-functionally unsatisfiable. Thus R is an associate of S.

10. We are given that M is maximally Γ-consistent and E-complete, and we are to show that M is a Hintikka set. In what follows, "consistent" will mean Γ-consistent. We shall say that a sentence X is consistent with M if the set $M \cup \{X\}$ is consistent. Since M is *maximally* consistent, then if X is consistent with M, then X must be a member of M (otherwise M would be only a *proper* subset of the larger consistent set $M \cup \{X\}$, contrary to M's maximality).

H_0: No formula and its conjugate (or formula and its negation, if unsigned formulas are in questions) can be in the set M, because M is consistent, and that is one of the conditions on (Γ) consistency.

H_1: Suppose $\alpha \in M$. Then of course α is consistent with M. Thus α_1 and α_2 are both consistent with M (since Γ is an analytic consistency property). Therefore α_1 and α_2 are members of M. This proves that if $\alpha \in M$, then $\alpha_1 \in M$ and $\alpha_2 \in M$.

H_2: Suppose $\beta \in M$. Then β is consistent with M and thus either β_1 is consistent with M or β_2 is consistent with M, and therefore either $\beta_1 \in M$ or $\beta_2 \in M$.

H_3: Suppose $\gamma \in M$. Then γ is consistent with M, so that $\gamma(a)$ is consistent with M for every parameter a. Therefore $\gamma(a) \in M$ for every parameter a.

H_4: Suppose $\delta \in M$. Then $\delta(a) \in M$ for at least one parameter a by the assumption that M is E-complete.

By H_1-H_4, M is a Hintikka set.

Part II
Recursion Theory and Metamathematics

Chapter 3

Some Special Topics

The items of this chapter represent reconstructions and generalizations of various results in recursion theory, undecidability and incompleteness.

First for a preliminary matter, which I should have addressed much earlier: By a *pair* (x, y) [more often called an *ordered pair*], is meant a set whose elements are x and y, together with a *designation* which specifies that x is the first element and y the second. In contrast, the *unordered pair* $\{x, y\}$ [note the curly brackets instead of parentheses] is just the set whose members are x and y, without any specification that one of the two elements comes first and the other second. The set $\{x, y\}$ is the same as the set $\{y, x\}$, but the pair (x, y) is different from the pair (y, x) [except in the case when $x = y$].

Throughout this chapter, the word *number* shall mean positive integer, and we will consider an operation that assigns to each pair (x, y) of numbers a number denoted xy, or sometimes $x * y$. We will assume the operator is such that xy (or $x * y$) uniquely determines the x and the y — that is, the only way that $x * y$ can be the same as $z * v$ is when $x = z$ and $y = v$. And we assume that every number n is assigned to some pair $x * y$, for some x and some y. We generally assume also that since the function $*$ is 1-1 and onto from the set of ordered pairs of positive integers to the set of positive integers, we can use its *inverse* to go from a given a number n to the pair (x, y) such that $n = x * y$.

I. A Decision Machine

We consider a calculating machine with infinitely many registers $R_1, R_2, \ldots,$ $R_n, \ldots$. We call n the *index* of R_n. Each register is, so to speak, in charge of a certain property of numbers. To find out whether a given number n has a certain property, one goes to the register in charge of that property and feeds in the number. The machine then goes into operation, and one of three things happens:

(1) The machine eventually halts and flashes a signal — say a green light — signifying that the number does have the property, in which case we say that the register *affirms n*.

(2) The machine eventually halts and flashes a different signal — say a red light — to indicate that the number doesn't have the property, in which case we say that the register *denies n*.

(3) The machine runs on forever, without ever being able to determine whether or not the number does have the property, in which case we say that n *stumps* the register, or that the register is *stumped* by n.

We assume we are given a machine M that satisfies the following two conditions:

M_1: With each register R in the calculating machine is associated another register R' in the calculating machine called the *opposer* of R, which affirms those numbers which R denies, and denies those numbers which R affirms (and hence is stumped by those and only those numbers which stump R).

M_2: With each register R in the calculating machine is associated another register $R^{\#}$ in the calculating machine called the *diagonalizer* of R, such that for any number x, the register $R^{\#}$ affirms x iff R affirms xx and denies x iff R denies xx. (I.e. $R^{\#}$ affirms x iff R affirms $x * x$ and $R^{\#}$ denies x iff R denies $x * x$, for our special function $*$ which has the properties listed above for it; we will write xy for $x * y$ from now on in this chapter.)

Recall our denumerable list of registers $R_1, R_2, \ldots, R_n, \ldots$. For any number n, we let n' be the index of $(R_n)'$, and let $n^{\#}$ be the index of $(R_n)^{\#}$. Thus $R_{n'} = (R_n)'$ and $R_{n\#} = (R_n)^{\#}$.

We will say that two registers R_a and R_b behave the same way towards a number n when they either both affirm n, or both deny n or are both stumped by n. We call two registers *similar* if they behave the same way towards every number. It is easy to see that for any register R, the registers $R'^{\#}$ and $R^{\#'}$ are similar, for each denies x iff R affirms xx, and each affirms x iff R denies xx. ($R^{\#'}$ means the same as $(R')^{\#}$ and $R^{\#'}$ means the same as

$(R^{\#})'$, but we will always eliminate the parentheses if the resulting meaning is clear.)

Universal Registers

We shall call a register U *universal* if, for all numbers x and y, the register U affirms xy iff R_x affirms y.

The propositions that follow will be proved later on in a more general framework. Meanwhile, some of you might like to try proving them on your own.

Proposition 1.1. *Any universal register can be stumped (by some number or other).*

Proposition 1.2. *If at least one of the registers is universal, then some register is stumped by its own index (i.e. for some number n, the register R_n is stumped by n).*

Contra-Universal Registers

We shall call a register V *contra-universal* if for all numbers x and y, the register V affirms xy iff R_x denies y.

Proposition 1.3. *Any contra-universal register V can be stumped.*

Proposition 1.4. *If some register is universal, then some register is stumped by the index of its opposer. If some register is contra-universal, then some register is stumped by its own index.*

Creative Registers

We shall call a register C *creative* if, for every register R, there is at least one number n such that C affirms n iff R affirms n.

Proposition 1.5. *Any creative register can be stumped.*

Proposition 1.6. *Every universal register is creative, and every contra-universal register is creative.*

Note: Propositions 1.5 and 1.6 provide additional and particularly simple proofs of Propositions 1.1 and 1.3.

For the rest of this section, we let A be the set of all numbers xy such that R_x affirms y, and let B be the set of all numbers xy such that R_x denies y.

Fixed Points

We call a number xy a *fixed point* of a register R if R affirms xy iff $xy \in A$ (i.e. R_x affirms y) and R denies xy iff $xy \in B$ (i.e. R_x denies y).

Proposition 1.7. *Every register has a fixed point.*

Proposition 1.8. *Suppose that U is universal and V is contra-universal. Then:*
(a) *Any fixed point of U' will stump U.*
(b) *Any fixed point of V will stump V.*

Proposition 1.9. *Suppose R is a register that affirms all numbers in A and doesn't affirm any number in B. Then R can be stumped.* [This proposition is particularly interesting.]

A Halting Problem

Let us say that a register R *halts* at a number n if R either affirms n or denies n. Call a register R a *stump detector* if R halts at those and only those numbers xy such that R_x is stumped by y.

Proposition 1.10. *No register is a stump detector.*

Domination

We say that n *dominates* m if R_n affirms all numbers affirmed by R_m.

Proposition 1.11. *Suppose U is universal and V is contra-universal and that R affirms all numbers affirmed by U and denies all numbers affirmed by V. Then R can be stumped.*

This proposition can be restated in terms of domination as follows: Suppose U is universal and V is contra-universal and there is a register R

such that R dominates U and R' dominates V. Then there exists an x such that R is stumped by x.

Affirmation Sets

By the *affirmation set* of a register R we shall mean the set of all numbers affirmed by R. We shall call a set of numbers S an *affirmation set* if it is the affirmation set of some register.

Proposition 1.12. *If one of the registers is universal, then there exists an affirmation set whose complement is not an affirmation set.*

II.   Variations on a Theme of Gödel

Some of the items here are repetitions of results in *The Beginner's Guide* [Smullyan, 2014]. The reason for these repetitions will be apparent later on. Again, the propositions that follow will be proved later, in a more general setting.

We consider a mathematical system $\mathcal{S}$ in which we have an infinite sequence $H_1, H_2, \ldots, H_k, \ldots$, of expressions called *predicates*, and to each predicate H and each number n is assigned an expression denoted $H(n)$, called a *sentence*. [Informally, we think of H as the name of some property of numbers, and of the sentence $H(n)$ as expressing the proposition that the number n has the property named by H.]

We assume that every sentence is $H(n)$ for some H and some n. We arrange all the sentences in a sequence $S_1, S_2, \ldots, S_k, \ldots$ in such a manner that for numbers x and y the sentence S_{x*y} is the sentence $H_x(y)$. We call n the *index* of the predicate H_n, and we call k the *index* of the sentence S_k. Thus $x * y$ is the index of the sentence $H_x(y)$.

Provable, Refutable and Undecidable Sentences

Some of the sentences are called *provable* and some are called *refutable*, and we assume the system is *consistent*, by which we mean that no sentence is both provable and refutable. A sentence is called *decidable* if it is either provable or refutable, and it is called *undecidable* if it is neither provable nor refutable. The system is called *complete* if every sentence is decidable, and is called *incomplete* otherwise. We let P be the set of all provable sentences and R be the set of all refutable sentences.

For any set W of sentences, we let W_0 be the set of all indices of the elements of W. Then P_0 is the set of all n such that S_n is provable, and R_0 is the set of all n such that S_n is refutable.

We shall call two sentences X and Y *equivalent* if when one is provable, so is the other, and when one is refutable, so is the other (and hence, when one is undecidable, so is the other); in other words, they are either both provable, both refutable, or both undecidable.

We are given that the system $\mathcal{S}$ obeys two conditions:

(1) With each predicate H is associated a predicate H' called the *negation* of H such that for every number n the sentence $H'(n)$ is provable iff $H(n)$ is refutable, and is refutable iff $H(n)$ is provable.

(2) With each predicate H is associated a predicate $H^{\#}$ called the *diagonalizer* of H such that for each number n the sentence $H^{\#}(n)$ is equivalent to the sentence $H(n * n)$.

Set Representation

We shall say that a predicate H *represents* the set of all numbers n such that $H(n)$ is provable. Thus to say that H represents a number set A is to say that for all numbers n, the sentence $H(n)$ is provable iff $n \in A$. We call the set A *representable* if some predicate represents it.

Provability and Refutability Predicates

We shall call H a *provability predicate* if it represents the set P_0. Thus the statement that H is a provability predicate is equivalent to the condition that for every n the sentence $H(n)$ is provable iff S_n is provable.

We shall call K a *refutability predicate* if it represents the set R_0. Thus the statement that K is a refutability predicate is equivalent to the condition that for every n the sentence $K(n)$ is provable iff S_n is refutable.

Proposition 2.1. *If H is a provability predicate, then $H(n)$ is undecidable for some n.*

Proposition 2.2. *If one of the predicates is a provability predicate, then there is some n such that $H_n(n)$ is undecidable.*

Proposition 2.3. *If K is a refutability predicate, then $K(n)$ is undecidable for some n.*

Recall that n' is the index of the negation of H_n, so that $(H_{n'}) = (H_n)'$.

Proposition 2.4. *If some predicate is a provability predicate, then there is some n such that $H_n(n')$ is undecidable. If some predicate is a refutability predicate, then there is some n such that $H_n(n)$ is undecidable.*

Creative Predicates

We shall call a predicate K *creative* if for every predicate H there is at least one number n such that $H(n)$ is provable iff $K(n)$ is provable.

Proposition 2.5. *If K is creative, then $K(n)$ is undecidable for some n.*

Proposition 2.6. *Every provability predicate is creative, and every refutability predicate is creative.*

Fixed Points

We shall call a number n a *fixed point* of a predicate H if S_n is equivalent to $H(n)$. [This is the same as saying that a number xy is a *fixed point* of a predicate H if $H_x(y)$ is equivalent to $H(xy)$.]

Proposition 2.7. *Every predicate has a fixed point.*

Proposition 2.8. *Suppose that H is a provability predicate and that K is a refutability predicate. Then:*
(a) *If n is a fixed point of H', then $H(n)$ is undecidable.*
(b) *If n is a fixed point of K, then $K(n)$ is undecidable.*

Proposition 2.9. *Suppose H is a predicate such that $H(n)$ is provable for any n in the set P_0, and $H(n)$ is not provable for any n in R_0. Then $H(n)$ is undecidable for some n.* [This is related to Rosser's incompleteness proof.]

Proposition 2.10. *There cannot be a predicate H such that for all numbers n the sentence $H(n)$ is decidable iff S_n is undecidable.*

Domination

We will say that a predicate H_b *dominates* a predicate H_a if $H_b(n)$ is provable for every n for which $H_a(n)$ is provable.

Proposition 2.11. *If H dominates some provability predicate, and H' dominates some refutability predicate, then $H(n)$ is undecidable for some n.*

Proposition 2.12. *If one of the predicates is a provability predicate, then there exists a representable number set whose complement is not representable.*

III.   *R*-Systems

The systems of I and those of II are really the same, only in different dress. The common underlying theme will be fully explained in the next section of this chapter. Meanwhile, we will consider yet another embodiment of the theme, which has direct applications to recursion theory.

We consider a denumerable collection of number sets, which we will call "R-sets", and an enumeration $A_1, A_2, \ldots, A_n, \ldots$ of them in some order, and another enumeration $B_1, B_2, \ldots, B_n, \ldots$ of them in some other order, such that for any disjoint pair (S_1, S_2) of R-sets, there is some n such that $S_1 = A_n$ and $S_2 = B_n$. We shall call such a pair of enumerations an *R-system* if the following three conditions hold:

R_0: For every n, the set A_n is disjoint from B_n i.e., $A_n \cap B_n = \emptyset$.

R_1: Associated with each number n is a number n' such that $A_{n'} = B_n$ and $B_{n'} = A_n$ [thus the ordered pair $(A_{n'}, B_{n'})$ is the pair (B_n, A_n)].

R_2: Associated with each n is a number $n^{\#}$ such that for every number x,
   (1) $x \in A_{n\#}$ iff $xx \in A_n$.
   (2) $x \in B_{n\#}$ iff $xx \in B_n$.

Note: The collection of all recursively enumerable sets can be arranged in a pair of sequences $A_1, A_2, \ldots, A_n, \ldots$ and $B_1, B_2, \ldots, B_n, \ldots$ such that conditions R_0, R_1 and R_2 all hold (for a *recursive* function $f(x, y) = x * y$). Thus all the results that follow are generalizations of results in recursion theory.

Universal and Contra-Universal Sets

We call an R-set A_w *universal* if for all numbers x and y the number $xy \in A_w$ iff $y \in A_x$. We call an R-set A_v *contra-universal* if for all numbers x and y the number $xy \in A_v$ iff $y \in B_x$.

Proposition 3.1. *If some R-set A_w is universal, then A_w is not the complement of B_w (i.e. some number x lies outside both A_w and B_w).*

Proposition 3.2. *If some R-set A_w is universal, then there is some n such that n itself lies outside both A_n and B_n.*

Proposition 3.3. *If the R-set A_v is contra-universal, then A_v is not the complement of B_v (i.e. some number x lies outside both A_v and B_v).*

Proposition 3.4. *If some R-set A_w is universal, then there is a number n such that n' lies outside both A_n and B_n. If some R-set A_v is contra-universal, then there is a number n such that n lies outside both A_n and B_n.*

Creative R-Sets

We shall call an R-set A_c *creative* if for every R-set A_x there is a least one number n such that $n \in A_c$ iff $n \in A_x$.

Proposition 3.5. *If A_c is creative, then A_c is not the complement of B_c.*

Proposition 3.6. *Any universal or contra-universal R-set is creative.*

Fixed Points

We now let A be the set of all numbers xy such that $y \in A_x$, and let B be the set of all numbers xy such that $y \in B_x$.

We shall call a number x a *fixed point* of the pair (A_n, B_n) if $x \in A_n$ iff $x \in A$ and $x \in B_n$ iff $x \in B$. [An equivalent definition is that a number xy is a *fixed point* of a pair (A_n, B_n) if $xy \in A_n$ iff $y \in A_x$ and $xy \in B_n$ iff $y \in B_x$.]

Proposition 3.7. *Every pair (A_n, B_n) has a fixed point.*

Proposition 3.8. (a) *If A_w is a universal R-set, then any fixed point of $(A_{w'}, B_{w'}) = (B_w, A_w)$ lies outside both A_w and B_w.* (b) *If A_v is a contra-universal R-set, then any fixed point of (A_v, B_v) lies outside both A_v and B_v.*

Proposition 3.9. *If $A \subseteq A_h$ and A_h is disjoint from B, then some number lies outside both A_h and B_h.*

Proposition 3.10. *There is no number n such that for all numbers x and y, $xy \in A_n \cup B_n$ iff $y \notin A_x \cup B_x$.*

R-Separable R-Sets

We shall say that an R-set S_1 is *R-separable* from an R-set S_2 if there is some R-set S such that S_1 is a subset of S and S_2 is a subset of the complement $\overline{S}$ of S.

Domination

We say that n *dominates* m when $A_m \subseteq A_n$, i.e. when A_m is a subset of A_n.

Proposition 3.11. *If A_w is universal and A_v is contra-universal, then A_w is not R-separable from A_v.* [This generalizes an important result in recursion theory, namely that there exist two disjoint recursively enumerable sets that are not recursively separable.]

Note. Proposition 3.11 *can* be rewritten in terms of domination, but the rewritten version is very long. It is simpler just to say that the solution to Problem 11 in the synthesis section (whose statement is expressed in terms of the concept of domination) can be used to prove Proposition 3.11 (by contradiction), as the reader will be able to see in the solutions to the problems of this chapter.

Proposition 3.12. *If some R-set is universal, then there is an R-set whose complement is not an R-set.* [This generalizes the important fact that there is a recursively enumerable set whose complement is not recursively enumerable.]

IV. A Synthesis

The Synthesis System

As already mentioned, the discussions and problems of sections I, II and III are really all the same, only dressed differently. In each of the three, we have numerical relations $A(x, y)$ and $B(x, y)$ such that the following three conditions hold:

D_0: $A(x, y)$ and $B(x, y)$ cannot both hold simultaneously for any ordered pair of numbers (x, y).

D_1: Associated with each number x is a number x' such that for all numbers y:
  (1) $A(x', y)$ iff $B(x, y)$;
  (2) $B(x', y)$ iff $A(x, y)$.

D_2: Corresponding to each number x is a number $x^{\#}$ such that for all y:
  (1) $A(x^{\#}, y)$ iff $A(x, y * y)$;
  (2) $B(x^{\#}, y)$ iff $B(x, y * y)$.

- For the decision machine M of I, $A(x, y)$ is the relation "R_x affirms y", and $B(x, y)$ is the relation "R_x denies y".
- For the mathematical system $\mathcal{S}$ of II, $A(x, y)$ is the relation "$H_x(y)$ is provable", and $B(x, y)$ is the relation "$H_x(y)$ is refutable".
- For the mathematical R-systems of III, $A(x, y)$ is the relation "$y \in A_x$", and $B(x, y)$ is the relation "$y \in B_x$".

We shall call the pair $A(x, y)$ and $B(x, y)$ a *duo* if conditions D_0, D_1 and D_2 hold. Any result proved about duos in general is simultaneously applicable to the items of sections I, II, and III. Thus, in the problems that follow, for each $n \leq 12$, the solution to Problem n simultaneously proves Propositions $1n, 2n$, and $3n$ of sections I, II, and III.

In this section we will again abbreviate $x * y$ by xy and we will omit the word "section" in front of I, II, and III. We let A be the set of all numbers xy such that $A(x, y)$ is true, and we let B be the set of all numbers xy such that $B(x, y)$ is true. It will be valuable to know that not only are $A(x, y)$ and $B(x, y)$ disjoint [for every pair (x, y)], but A and B are also disjoint (contain no common element n). This is because every n is xy for some

x and y, and it is given that $A(x,y)$ and $B(x,y)$ don't both hold, which means that xy cannot be in both A and B, and thus n cannot be in both A and B.

- For the decision machine M of I, A is the set of all numbers xy such that R_x affirms y, and B is the set of all numbers xy such that R_x denies y.
- For the mathematical system S of II, A is the set P_0 (which is the set of all numbers xy such that $H_x(y)$ is provable), and B is the set R_0 (the set of all numbers xy such that $H_x(y)$ is refutable).
- For the mathematical R-systems of III, A is the set of all numbers xy such that $y \in A_x$, and B is the set of all numbers xy such that $y \in B_x$.

Undecidable Numbers

We shall call a number n *undecidable* if n is in neither A nor B. Thus xy is undecidable if neither $A(x,y)$ nor $B(x,y)$ holds. If the number n is not *undecidable*, we call it *decidable*.

- For the decision machine M of I, $n = xy$ being undecidable means that R_x is stumped by y.
- For the mathematical system S of II, $n = xy$ being undecidable means that the sentence $H_x(y)$ is undecidable (i.e. neither provable nor refutable).
- For the mathematical R-systems of III, $n = xy$ being undecidable means that y lies outside both A_x and B_x.

The following is an important fact: Since A is disjoint from B, it follows that to prove that a number n is undecidable, it suffices to show that $n \in A$ iff $n \in B$, because this would mean that n is either in both A and B, or in neither one. Since it cannot be in both, it must be in neither one, i.e. it is undecidable.

Universal Numbers

We shall call a number w *universal* if for all x it is the case that $wx \in A$ iff $x \in A$, or, which is the same thing, for all x and y, the relation $A(w, xy)$ holds iff $A(x,y)$ holds.

- For the decision machine M of I, to say that R_w is a universal register is to say that for all x and y, R_w affirms xy iff R_x affirms y.
- For the mathematical system S of II, to say that H_w is a provability predicate (a "universal predicate for provability") is to say that for all x

and y, $H_w(xy)$ is provable iff S_{xy} is provable (which is the same as saying $H_x(y)$ is provable).

- For the mathematical R-systems of III, to say that w is universal is to say that A_w is a universal set, i.e. that $xy \in A_w$ iff $y \in A_x$.

Contra-Universal Numbers

We call a number v *contra-universal* if $vx \in A$ iff $x \in B$, or, what is the same thing, for all x and y, $A(v, xy)$ iff $B(x, y)$.

- For the decision machine M of I, to say that R_v is a contra-universal register is to say that for all x and y, R_v affirms xy iff R_x denies y.
- For the mathematical system S of II, to say that H_v is a refutability predicate (a "contra-universal predicate") is to say that for all x and y, $H_v(xy)$ is provable iff S_{xy} is refutable (which is the same as saying that $H_x(y)$ is refutable or that $\sim H_x(y)$ is provable).
- For the mathematical R-systems of III, to say that v is contra-universal is to say that, for all x and y, $xy \in A_v$ iff $y \in B_x$.

Creative Numbers

We call a number c *creative* if for every number n there is at least one number x such that $cx \in A$ iff $nx \in A$.

- For the decision machine M of I, to say that a register R_c is creative is to say that for every register R_n, there is at least one number x such that R_c affirms x iff R_n affirms x.
- For the mathematical system S of II, to say the predicate H_c is creative is to say that for every predicate R_n, there is at least one number x such that $H_c(x)$ is provable iff $H_n(x)$ is provable.
- For the mathematical R-systems of III, to say that c is creative is to say that, for every number n, there is at least one number x such that $x \in A_c$ iff $x \in A_n$.

Fixed Points

We let "x is a *fixed point* of n" mean that $nx \in A$ iff $x \in A$ and $nx \in B$ iff $x \in B$.

- For the decision machine M of I, to say that a number $x * y$ is a *fixed point* of a register R is to say that R affirms $x * y$ iff $x * y \in A$ (i.e. R_x affirms y) and R denies $x * y$ iff $x * y \in B$ (i.e. R_x denies y).

- For the mathematical system S of II, to say that xy is a fixed point of the predicate H_n is to say that $H_n(xy)$ is equivalent to S_{xy}, i.e. both these sentences are provable, or both are refutable, or both are neither. [This is the same as saying that $H_n(xy)$ is equivalent to $H_x(y)$.]
- For the mathematical R-systems of III, to say that a number x is a *fixed point* of a pair (A_n, B_n) if $x \in A_n$ iff $x \in A$ and $x \in B_n$ iff $x \in B$.

Problem 1. Prove that if w is universal, then there is at least one number x such that wx is undecidable.

Problem 2. Prove that if some number is universal, then there is at least one number x such that xx is undecidable.

Problem 3. Prove that if v is contra-universal, then there is at least one number x such that vx is undecidable.

Problem 4. Prove that if some number is universal, then there is some number x such that xx' is undecidable. And prove that if some number is contra-universal, then there is some number x such that xx is undecidable.

Problem 5. Prove that if c is creative, then there is at least one number x such that cx is undecidable.

Problem 6. Prove that every universal number is creative, and that every contra-universal number is creative.

Problem 7. Prove that every number has a fixed point.

Problem 8. Suppose w is universal and v is contra-universal. Prove:
(a) If x is any fixed point of w', then x is undecidable, and so is wx.
(b) If x is any fixed point of v, then x is undecidable, and so is vx.

Problem 9. Suppose h is a number such that, for all numbers x:
(a) If $x \in A$, then $hx \in A$.
(b) If $x \in B$, then $hx \notin A$.
Prove that hx is undecidable for some x.

Why is this a strengthening of the result of Problem 1?

Problem 10. Prove that there is no number n such that for all numbers x the number nx is decidable if and only if x is undecidable.

Domination

Let us say that a number n *dominates* a number m if, for every number x, if $mx \in A$, then $nx \in A$.

- For the decision machine M of I, to say that the register R_n dominates the register R_m is to say that R_n affirms all numbers affirmed by R_m.
- For the mathematical system S of II, to say that the predicate H_n dominates the predicate H_m is to say that that for all x, if $H_m(x)$ is provable, so is $H_n(x)$.
- For the mathematical R-systems of III, to say that A_n dominates A_m is to say that $A_m \subseteq A_n$.

Problem 11. Show that if n dominates some universal number w and n' dominates some contra-universal number v, then nx is undecidable for some x.

Set Representation

We will say that the number n *represents* the set S if, for all x, $nx \in A$ iff $x \in S$.

- For the decision machines of I, to say that a register R represents a set S is just to say that the set S is the affirmation set of the register R.
- For the mathematical systems S of II, to say that predicate H represents a set S is just to say that S is the set of all numbers n such that $H(n)$ is provable.
- For the mathematical R-systems of III, to say that the R-set A_n represents a set S is just to say that $y \in A_n$ iff $y \in S$, i.e., $A_n = S$.

Problem 12. Prove that if some number is universal, then there is a representable set whose complement is not representable.

Solutions to the Problems of Chapter 3

1. Suppose w is universal. Then for any number x, $w(w^{\#'}x) \in A$ holds iff $w^{\#'}x \in A$, which is true iff $w^{\#}x \in B$, which is true iff $w(xx) \in B$. Thus:

(1) $w(w^{\#'}x) \in A$ iff $w(xx) \in B$.

If we take $w^{\#'}$ for x in (1), we obtain:

(2) $w(w^{\#'}w^{\#'}) \in A$ iff $w(w^{\#'}w^{\#'}) \in B$.

But A and B are disjoint, so $w(w^{\#'}w^{\#'})$ can be in neither A nor B, which shows that $w(w^{\#'}w^{\#'})$ is undecidable; $w(w'^{\#}w'^{\#})$ is also undecidable, as the reader can verify.

2. Suppose w is universal. Then, by D_1, for any number x, $w^{\#'}x \in B$ holds iff $w^{\#}x \in A$, which is true iff $w(xx) \in A$, which is true iff $xx \in A$. Thus $w^{\#'}x \in B$ iff $xx \in A$. We take $w^{\#'}$ for x and obtain $w^{\#'}w^{\#'} \in B$ iff $w^{\#'}w^{\#'} \in A$. Hence $w^{\#'}w^{\#'}$ is undecidable.

3. Suppose v is contra-universal. Then for any number x, $v(v^{\#}x) \in A$ is true iff $v^{\#}x \in B$, which is true iff $v(xx) \in B$. Thus $v(v^{\#}x) \in A$ iff $v(xx) \in B$. We take $v^{\#}$ for x and obtain $v(v^{\#}v^{\#}) \in A$ iff $v(v^{\#}v^{\#}) \in B$. Thus $v(v^{\#}v^{\#})$ is undecidable.

4. Suppose w is universal. We have seen in the solution to Problem 2 that $w^{\#'}w^{\#'}$ is undecidable. Therefore $w^{\#}w^{\#'}$ is undecidable. [For any numbers x and y, $x'y$ is undecidable iff xy is undecidable. (Why?)] Thus nn' is undecidable for $n = w^{\#}$.

   Next, suppose that v is contra-universal. For any number x, we know that $v^{\#}x \in A$ iff $v(xx) \in A$. We take $v^{\#}$ for x and obtain $v^{\#}v^{\#} \in A$ iff $v(v^{\#}v^{\#}) \in A$, which in turn is the case iff $v^{\#}v^{\#} \in B$ (since v is contra-universal). Thus $v^{\#}v^{\#} \in A$ iff $v^{\#}v^{\#} \in B$, so that $v^{\#}v^{\#}$ must be undecidable.

5. Suppose c is creative. Then for any x there is some n such that $cn \in A$ iff $xn \in A$. Taking c' for x, we obtain that for some n, $cn \in A$ iff $c'n \in A$. But $c'n \in A$ iff $cn \in B$. Thus $cn \in A$ iff $cn \in B$, so that cn is undecidable.

6. (a) Suppose w is universal. For any number x, $w(x^{\#}x^{\#}) \in A$ is true iff $x^{\#}x^{\#} \in A$, which is true iff $x(x^{\#}x^{\#}) \in A$. Thus $wn \in A$ iff $xn \in A$, for $n = x^{\#}x^{\#}$.

   (b) Suppose v is contra-universal. For any x, $v(x^{\#'}x^{\#'}) \in A$ is true iff $x^{\#'}x^{\#'} \in B$, which is true iff $x^{\#}x^{\#'} \in A$, which is true iff $x(x^{\#'}x^{\#'}) \in A$. Thus $vn \in A$ iff $xn \in A$, for $n = x^{\#'}x^{\#'}$.

7. For any number n, $n^{\#}n^{\#} \in A$ holds iff $n(n^{\#}n^{\#}) \in A$, and $n^{\#}n^{\#} \in B$ holds iff $n(n^{\#}n^{\#}) \in B$. Hence $n^{\#}n^{\#}$ is a fixed point of n.

8. (a) Suppose w is universal and n is a fixed point of w'. Then $n \in A$ iff $wn \in A$ is true (since w is universal), which is true iff $w'n \in B$ (by D_1), which is true iff $n \in B$ (since n is a fixed point of w'). Thus $n \in A$ iff $n \in B$, so that n is undecidable. Since n is a fixed point of w', then $w'n$ is also undecidable (why?), and hence wn is undecidable.

   (b) Suppose v is contra-universal and n is a fixed point of v. Then $n \in A$ iff $vn \in A$ is true (since n is a fixed point of v). But $vn \in A$ iff $n \in B$ (since v is contra-universal). Thus $n \in A$ iff $n \in B$, and is thus undecidable. Since n is a fixed point of v, it is also true that $vn \in B$ iff $n \in B$. Since n can be in neither A nor B, it follows from what we just saw that vn can be in neither A nor B either, and is thus also undecidable.

Remarks. We know that $w'^{\#}w'^{\#}$ is a fixed point of w' (since by the solution to Problem 7, $x^{\#}x^{\#}$ is a fixed point of x, for any x). But it is also true that $w^{\#'}w^{\#'}$ is a fixed point of w' (in fact $x^{\#'}x^{\#'}$ is a fixed point of x', for any x, as the reader can verify). Thus (a) alone provides another proof that $w(w^{\#'}w^{\#'})$ is undecidable.

9. We are given that for all x:

   $$\text{If } x \in A, \text{ then } hx \in A.$$
   $$\text{If } x \in B, \text{then } hx \notin A.$$

   Let k be the number $h^{\#'}$. We first show that kk is undecidable.

   (a) Suppose $kk \in A$. Then $h(kk) \in A$ [by (a)]. Thus $h^{\#}k \in A$ so that $h^{\#'}k \in B$. And thus $kk \in B$ [since $k = h^{\#'}$]. Thus if $kk \in A$ it is also true that $kk \in B$, but kk cannot be in both A and B, and so $kk \notin A$.

   (b) Suppose $kk \in B$. Then $h(kk) \notin A$ [by (b)]. Hence $h^{\#'}k \notin B$, which means that $kk \notin B$ [since $k = h^{\#'}$]. Thus if $kk \in B$, then $kk \notin B$, which is a contradiction. Therefore $kk \notin B$.

   Thus $kk \notin A$ and $kk \notin B$, and so kk is undecidable. Thus $h^{\#'}h^{\#'}$ is undecidable, hence so is $h^{\#}h^{\#'}$, and therefore so is $h(h^{\#'}h^{\#'})$.

   Note: The above proof is from scratch. A swifter proof is possible using the fact that every number has a fixed point. We leave it to the reader to show that if n is any fixed point of h, then hn is undecidable.

   Next, why is this result a strengthening of the result of Problem 1? The answer is that if h is universal, then it obviously satisfies conditions (a) and (b).

10. If a number n exists, it must have a fixed point, by Problem 7. But if
    x is a fixed point of n, then (by the definition of a fixed point) $nx \in A$
    iff $x \in A$ and $nx \in B$ iff $x \in B$. Clearly, for such an x, it cannot be
    that nx if decidable iff x is undecidable.

11. This follows easily from Problem 9: Suppose h dominates w and h'
    dominates v, where w is universal and v is contra-universal.

    If $x \in A$, then $wx \in A$, so that $hx \in A$ [since h dominates w]. Thus,
    $$(1) \text{ If } x \in A, \text{ then } hx \in A.$$

    Next, suppose $x \in B$. Then $vx \in A$, so that $h'x \in A$, and then
    $hx \in B$, and finally $hx \notin A$. Thus,
    $$(2) \text{ If } x \in B, \text{ then } hx \notin A.$$

    Thus, by (1) and (2), h satisfies the hypotheses of Problem 9, and so
    hx is undecidable for some x.

    Because the relationship between the solution to Problem 11 and
    Propositions 11 and 35 are a little more complex than some of the
    others, we will give the proofs of these propositions (using the solution
    to Problem 11) here. First let us review what "n dominates m" means
    for the first three sections of this chapter:

    I: R_n affirms all numbers affirmed by R_m.

    II: For all x, if $H_m(x)$ is provable, so is $H_n(x)$.

    III: $A_m \subseteq A_n$.

    First let us consider Proposition 11: Suppose U is universal and V
    is contra-universal and that R affirms all numbers affirmed by U and
    denies all numbers affirmed by V. Then R can be stumped.

    In the text we gave a restatement of this claim in terms of domination:

    > Suppose U is universal and V is contra-universal and there is a
    > register R such that R dominates U and R' dominates V. Then
    > there exists an x such that R is stumped by x.

    It is easy to see that the restatement is equivalent to the original
    statement, for to say that R denies all numbers affirmed by V is equiv-
    alent to saying that R' affirms all numbers affirmed by V. To bring this
    restatement closer to the expression of Problem 11 in the synthesis, all
    we have to do is the following: We let n be the index of R, w be the
    index of U, and v be the index of V; then the restatement of Proposi-
    tion 11 is equivalent to: If R_n dominates the universal register R_w and
    R'_n dominates the contra-universal register R_v, then the register R_n

can be stumped by some x. And this is the statement of Problem 11, reinterpreted in the vocabulary of the concepts of I.

The application to Proposition 23 is obvious.

Finally, let us consider Proposition 35: If A_w is universal and A_v is contra-universal, then A_w is not R-separable from A_v.

To show that A_w is not R-separable from A_v, suppose that S_1 and S_2 are disjoint R-sets such that $A_w \subseteq S_1$ and $A_v \subseteq S_2$. We must show that some number lies outside of both S_1 and S_2. Now, for some number n, $S_1 = A_n$ and $S_2 = B_n$. Then $S_1 = A_n$ and $S_2 = A_{n'}$. So $A_w \subseteq A_n$ and $A_v \subseteq A_{n'}$, which means that n dominates w and n' dominates v, and so, by the solution to Problem 11, there is some number x such that nx is undecidable, i.e $x \notin A_n$ and $x \notin B_n$. Thus $x \notin A_n$ and $x \notin A_{n'}$, i.e. $x \notin S_1$ and $x \notin S_2$, which is what was to be shown.

12. We first show that no number x can represent the complement $\overline{A}$ of A. To show this it suffices to demonstrate that, given any number x there is at least one number y such that it is *not* the case that $xy \in A$ iff $y \in \overline{A}$, or, what is the same thing, that there is at least one number y such that it *is* the case that $xy \in A$ iff $y \in A$. Well, any fixed point y of x is such a number. Thus the set $\overline{A}$ is not representable.

Now, if there is a universal number w, then w represents A (why?), so that A is then a representable set whose complement $\overline{A}$ is not representable.

Chapter 4

Elementary Formal Systems
and Recursive Enumerability

I. More on Elementary Formal Systems

Let us review the definition of an elementary formal system and some of the basic properties of such systems established in *The Beginner's Guide* [Smullyan, 2014].

By an *alphabet* K is meant a finite sequence of elements called *symbols*. Any finite sequence of these symbols is called a *string* or *expression* in K.

By an *elementary formal system* (E) over K is meant a collection of the following items:

(1) The alphabet K.

(2) Another alphabet V of symbols called *variables*. We will usually use the letters x, y, z, w with or without subscripts, as variables.

(3) Still another alphabet of symbols called *predicates*, each of which is assigned a positive integer called the *degree* of the predicate. We usually use capital letters for predicates.

(4) Two more symbols called the *punctuation sign* (usually a comma) and the *implication sign* (usually "→").

(5) A finite sequence of expressions, all of which are *formulas* (formulas will be defined below).

By a *term* is meant any string composed of symbols of K and variables. A term without variables is called a *constant term*. By an *atomic formula* is meant an expression Pt, where P is a predicate of degree 1 and t is a term, or an expression $Rt_1, \ldots, t_n$, where R is a predicate of degree n and $t_1, \ldots, t_n$ are terms.

By a *formula* is meant an atomic formula, or an expression of the form $F_1 \to F_2 \ldots \to F_n$, where $F_1, F_2, \ldots, F_n$ are atomic formulas. This concludes the definition of a formula.

By a *sentence* is meant a formula with no variables.

By an *instance* of a formula is meant the result of substituting *constant* terms for all variables in the formula.

An elementary formal system (E) over K is thus formed by distinguishing a particular finite subset of the formulas, which are called the *axiom schemes* of the system. The set of all instances of all the axiom schemes is called the set of *axioms of the system*.

A sentence is called *provable* in the elementary formal system (E) if its being so is a consequence of the following two conditions:
(1) Every axiom of the system is provable.
(2) For every *atomic* sentence X and every sentence Y, if X and $X \to Y$ are provable, so is Y.

More, explicitly, by a *proof* in the system is meant a finite sequence of sentences such that each member Y of the sequence is either an axiom of the system, or there is an *atomic* sentence X such that X and $X \to Y$ are earlier members of the sequence. A sentence is then called *provable* in the system if it is an element in the sequence of some proof in the system.

Elementary formal systems are designed to formalize the nature of a mechanical procedure, and the provable sentences of a given elementary formal system that have the form of some predicate P of degree n followed by an n-tuple of elements of the alphabet K are just those that can be generated by the given elementary formal system (which must of course include the predicate P in at least one of its axiom schemes if such a sentence is to be provable). And when such a term $Pk_1 \ldots k_n$ can be generated by the system, we say that it is *true* that the n-tuple $k_1, \ldots, k_n$ stands in the relation P, or we may say that $Pk_1 \ldots k_n$ is true. Thus we identify provability and truth when it comes to elementary formal systems.

Representability

We say that a predicate P of degree n *represents* a relation R of degree n of n-tuples of strings of K if for all constant terms $x_1, \ldots, x_n$, the sentence $Px_1, \ldots, x_n$ is provable in (E) iff the relation $Rx_1, \ldots, x_n$ holds. For $n = 1$, a predicate P of degree (1) represents the set of all constant terms x such that Px is provable in (E). A set or relation is said to be *representable* in (E) if some predicate represents it, and a relation or set is called *formally representable* if it is representable in some elementary formal system.

Some Basic Closure Properties

For any $n \geq 2$ we regard a relation of degree n on a set S as being a set of n-tuples $(x_1, \ldots, x_n)$ of members of S. We consider subsets of S to be special cases of relations on S, namely relations of one argument. We now wish to establish some basic closure properties of the collection of all relations (and sets) representable over K.

First, let us review some more facts discussed in *The Beginner's Guide* [Smullyan, 2014]. Consider two elementary formal systems (E_1) and (E_2) over a common alphabet K. We shall call the two systems *independent* if they contain no common predicates. By $(E_1) \cup (E_2)$ we shall mean the EFS (elementary formal system) whose axioms are those of (E_1) together with those of (E_2). Obviously every provable sentence of (E_1) and of (E_2) is provable in $(E_1) \cup (E_2)$. And if (E_1) and (E_2) are *independent*, then any sentence provable in $(E_1) \cup (E_2)$ is clearly provable in (E_1) *or* (E_2) alone, and hence the representable relations of $(E_1) \cup (E_2)$ are the representable relations in (E_1) alone, together with those representable in (E_2) alone.

Similarly, if $(E_1), (E_2), \ldots, (E_n)$ are mutually independent (no two having a common predicate), then the representable relations of $(E_1) \cup \ldots \cup (E_n)$, are the result of joining together those of each of $(E_1), \ldots, (E_n)$ alone.

Now suppose $R_1, \ldots, R_n$ are each formally representable over K. Since we are assuming an unlimited stack of predicates at our disposal, we can represent $R_1, \ldots, R_n$ in mutually independent systems $(E_1), (E_2), \ldots, (E_n)$, and so we have:

Proposition 1. *If $R_1, \ldots, R_n$ are each formally representable over K, then they can all be represented in a common EFS over K.*

Now for our closure properties:

(a) *Unions and Intersections.* For any two relations $R_1(x_1, \ldots, x_n)$, $R_2(x_1, \ldots, x_n)$ of the same degree, by their union $R_1 \cup R_2$ — also written $R_1(x_1, \ldots, x_n) \vee R_2(x_1, \ldots, x_n)$ — is meant the set of all n-tuples $(x_1, \ldots, x_n)$ that are in either R_1 or R_2 or in both. Thus $(R_1 \cup R_2)(x_1, \ldots, x_n)$ iff $R_1(x_1, \ldots, x_n) \vee R_2(x_1, \ldots, x_n)$. By the intersection $R_1 \cap R_2$ — also written $R_1(x_1, \ldots, x_n) \wedge R_2(x_1, \ldots, x_n)$ — is meant the relation that holds for $(x_1, \ldots, x_n)$ iff

$$R_1(x_1, \ldots, x_n) \wedge R_2(x_1, \ldots, x_n):$$

(b) *Existential Quantifiers.* For any relation $R(x_1, \ldots, x_n, y)$ of degree two or higher, by its existential quantification $\exists y R(x_1, \ldots, x_n, y)$ — also abbreviated $\exists R$ — is meant the set of all n-tuples $(x_1, \ldots, x_n)$ such that $R(x_1, \ldots, x_n, y)$ holds for at least one y:

(c) *Explicit Transformation.* Let n be any positive integer and consider n variables $x_1, \ldots, x_n$. Let R be a relation of degree k and let $\alpha_1, \ldots, \alpha_k$ each be either one of the n variables $x_1, \ldots, x_n$ or a constant term (i.e. a string of symbols of K). By the relation $\lambda x_1, \ldots, x_n R(\alpha_1, \ldots, \alpha_k)$ is meant the set of all n-tuples $(a_1, \ldots, a_n)$ of strings of K such that $R(b_1, \ldots, b_k)$ holds, where each b_i is defined as follows:

(1) If α_i is one of the variables x_j, then $b_i = a_j$.

(2) If α_i is a constant c, then $b_i = c$.

[For example, for $n = 3$ and constants c and d, $\lambda x_1, x_2, x_3 R(x_3, c, x_2, x_2, d)$ is the set of all triples (a_1, a_2, a_3) such that the relation $R(a_3, c, a_2, a_2, d)$ holds.]

In this situation, we say that the n-ary relation

$$\lambda x_1, \ldots, x_n R(\alpha_1, \ldots, \alpha_k)$$

is *explicitly definable* from the k-ary relation R and that the operation which assigns $\lambda x_1, \ldots, x_n R(\alpha_1, \ldots, \alpha_k)$ to the relation R is an *explicit transformation.*

Problem 1. Show that the collection of relations formally representable over K is closed under unions, intersections, existential quantification and explicit transformations. In other words, prove the following:

(a) If R_1 and R_2 are of the same degree and formally representable over K, then $R_1 \cup R_2$ and $R_1 \cap R_2$ are formally representable over K and of the same degree as R_1 and R_2.

(b) If the relation $R(x_1, \ldots, x_n, y)$ is a relation of degree $n + 1$ (n greater than or equal to one) that is formally representable over K, then the relation $\exists y R(x_1, \ldots, x_n, y)$ is a relation of degree n that is formally representable over K.

(c) If the relation $R(x_1, \ldots, x_k)$ is a relation of degree k that is formally representable over K, then $\lambda x_1, \ldots, x_n R(\alpha_1, \ldots, \alpha_k)$ is a relation of degree n that is formally representable over K (here each α_i is one of the variables $x_1, \ldots, x_n$ or a constant).

We recall that a set or relation W is said to be *solvable* over K if W and its complement $\overline{W}$ are both formally representable over K.

Problem 2. Prove that the collection of all sets and relations which are *solvable* over K is closed under union, intersection, complementation and explicit transformation.

II. Recursive Enumerability

We shall now identify the positive integers with the *dyadic numerals* which represent them (as defined in *The Beginner's Guide* [Smullyan, 2014]). We consider the two-sign alphabet $\{1, 2\}$, which we call "D", and define a *dyadic system* to be an elementary formal system over the alphabet D. We define a numerical relation to be *recursively enumerable* if it is representable in some dyadic system. (In *The Beginner's Guide* we called these relations *dyadically representable*.) It turns out that the recursively enumerable sets and relations are the same as the Σ_1-relations — and also the same as those representable in Peano Arithmetic.

We call a set R *recursive* if it and its complement $\overline{R}$ are both recursively enumerable. Thus a set is recursive if it is solvable in some dyadic system.

In *The Beginner's Guide*, we showed that the relations Sxy (y is the successor of x), $x < y, x \leq y, x = y, x \neq y, x + y = z, x \times y = z$ and $x^y = z$ are all recursively enumerable (in fact, recursive).

We also know (taking D as our alphabet K) that the collection of recursively enumerable relations is closed under unions, intersections, explicit transformations, and existential quantifications. We now need some further closure properties.

Finite Quantifications (Bounded Quantifiers)

For any numerical relation $R(x_1, \ldots, x_n, y)$, by the relation

$$(\forall z \leq y) R(x_1, \ldots, x_n, z)$$

[sometimes abbreviated by $\forall_F R$ and called the *finite universal quantification of R*] is meant the set of all $(n + 1)$-tuples $(x_1, \ldots, x_n, y)$ such that $R(x_1, \ldots, x_n, z)$ holds for every z that is less than or equal to y. Thus $(\forall z \leq y) R(x_1, \ldots, x_n, z)$ holds iff $R(x_1, \ldots, x_n, 1)$, $R(x_1, \ldots, x_n, 2), \ldots, R(x_1, \ldots, x_n, y)$ all hold. $(\forall z \leq y) R(x_1, \ldots, x_n, z)$ is equivalent to the formula

$$\forall z (z \leq y \supset R(x_1, \ldots, x_n, z)).$$

By the relation $(\exists z \leq y) R(x_1, \ldots, x_n, z)$ [sometimes abbreviated by $\exists_F R$ and called the *finite existential quantification of R*] is meant the set of all $(n + 1)$-tuples $(x_1, \ldots, x_n, y)$ such that there is at least one $z \leq y$ such that $R(x_1, \ldots, x_n, z)$ holds. Thus $(\exists z \leq y) R(x_1, \ldots, x_n, z)$ is the *disjunction* of the relations $R(x_1, \ldots, x_n, 1)$, $R(x_1, \ldots, x_n, 2), \ldots, R(x_1, \ldots, x_n, y)$. The formula $(\exists z \leq y) R(x_1, \ldots, x_n, z)$ is equivalent to the formula

$$\exists z (z \leq y \wedge R(x_1, \ldots, x_n, z)).$$

The quantifiers $(\forall x \leq c)$ and $(\exists x \leq c)$ are called *bounded quantifiers*.

Problem 3. Show the following:

(a) If $R(x_1, \ldots, x_n, y)$ is recursively enumerable, so are $\forall_F R$ and $\exists_F R$.
(b) If $R(x_1, \ldots, x_n, y)$ is recursive, so are $\forall_F R$ and $\exists_F R$.

Before continuing, let us recall the definitions of Σ_0 and Σ_1 formulas of elementary arithmetic. In *The Beginner's Guide* [Smullyan, 2014] we defined a Σ_0 formula and a Σ_0 relation as follows: By an *atomic Σ_0 formula* we mean any formula of one of the forms $c_1 = c_2$ or $c_1 \leq c_2$, where c_1 and c_2 are both terms. We then define the class of Σ_0 formulas of Elementary Arithmetic by the following inductive scheme:

(1) Every atomic Σ_0 formula is a Σ_0 formula.
(2) For any Σ_0 formulas F and G, the formulas $\sim F, F \wedge G, F \vee G$, and $F \supset G$ are Σ_0 formulas.
(3) For any Σ_0 formula F, and any variable x, and any c which is either a Peano numeral or a variable distinct from x, the expressions $(\forall x \leq c)F$ and $(\exists x \leq c)F$ are Σ_0 formulas.

No formula is a Σ_0 formula unless its being so is a consequence of (1), (2) and (3) above.

Thus a Σ_0 formula is a formula of Elementary Arithmetic in which all quantifiers are bounded. A set or relation is called a Σ_0 set or relation (or just "Σ_0") if it is expressed by a Σ_0 formula.

A relation is called Σ_1 if it is expressed by a formula of the form $\exists z R(x_1, \ldots, x_n, z)$, where the relation $R(x_1, \ldots, x_n, z)$ is Σ_0.

We noted in *The Beginner's Guide* [Smullyan, 2014] that all Σ_0 and Σ_1 relations and sets are *arithmetic*, because the formulas that express them are just a subset of all the formulas of Elementary Arithmetic that can be used to express number sets and relations. Moreover, we discussed the fact that, given any Σ_0 sentence (a Σ_0 formula with no free variables), we can effectively decide whether it is true or false.

By Problem 2, taking D for the alphabet K, the collection of all recursive sets and relations is closed under unions, intersections, complements, and explicit transformations. It is also closed under *finite* universal and existential quantifications, as we have just seen. This collection also contains the relations $x + y = z$ and $x \times y = z$. From all these facts, we have the following proposition:

Proposition 2. *Every Σ_0-relation is recursive.*

We also know by Problem 1, taking D for the alphabet K, that the existential quantification of a recursively enumerable relation is recursively enumerable. Now, every Σ_1-relation is the existential quantification of a Σ_0-relation, hence of a recursively enumerable (in fact recursive) relation, and is therefore recursively enumerable. This proves that:

Proposition 3. *All Σ_1-relations are recursively enumerable.*

We now wish to prove that, conversely, all recursively enumerable relations are Σ_1, and hence that being recursively enumerable is the same as being Σ_1!

First, for some very useful facts about Σ_1-relations:

Problem 4. Suppose $R(x_1, \ldots, x_n, y)$ is Σ_1. Show that the relation $\exists y R(x_1, \ldots, x_n, y)$ [as a relation between $x_1, \ldots, x_n$] is Σ_1.

For any function $f(x)$ (from numbers to numbers), we say that the function $f(x)$ is Σ_1 if the relation $f(x) = y$ is Σ_1.

Proposition 4. *If A is a Σ_1 set and $f(x)$ is a Σ_1-function, and A' is the set of all x such that $f(x) \in A$, then A' is Σ_1.* [A' is what is called the domain *of the function* $f(x)$ when the range of the function is restricted to the set A.]

Problem 5. Prove the above proposition.

To prove that every recursively enumerable relation is Σ_1, we will illustrate the proof for recursively enumerable *sets*. We will show that every recursively enumerable set A is Σ_1. The proof for relations is a modification we leave to the reader.

We use the dyadic Gödel numbering of *The Beginner's Guide* [Smullyan, 2014]: For any ordered alphabet K, viz. $\{k_1, k_2, \ldots, k_n\}$, we assign the Gödel number 12 to k_1, 122 to k_2, and a 1 followed by n 2's to k_n. And then, for any compound expression, we take its Gödel number to be the result of replacing each symbol of K by its Gödel number. We use g to denote the Gödel number function, e.g. the Gödel number of $g(k_3k_1k_2) = 122212122$.

In *The Beginner's Guide*, we showed that if a set S of strings of symbols of K is formally representable over K, then the set S_0 of Gödel numbers of the members of S is Σ_1. Of course this also holds when K is the two-sign alphabet $\{1, 2\}$. Thus if A is recursively enumerable, then the set A_0 of the Gödel numbers of the members of A (in dyadic notation) is Σ_1. Our job now is to get from the fact that a set of Gödel numbers A_0 is Σ_1 to the fact that A itself is Σ_1. Now, A is the set of all x such that $g(x) \in A_0$, and so [by Proposition 4 above] it suffices to show that the relation $g(x) = y$ is Σ_1. To this end, we use the following lemma, which was proved in *The Beginner's Guide*:

Lemma K. *There exists a Σ_0 relation $K(x, y, z)$ such that the following two conditions hold:*

(1) *For any finite set $(a_1, b_1), \ldots, (a_n, b_n)$ of ordered pairs of numbers (positive integers), there is a number z (called a* code number *of the set) such that for all x and y, $K(x, y, z)$ holds if and only if (x, y) is one of the pairs $(a_1, b_1), \ldots, (a_n, b_n)$.*

(2) *For all x, y, z if $K(x, y, z)$ holds, then $x < z$ and $y < z$*

Problem 6. Using Lemma K, prove the relation $g(x) = y$ is Σ_1.

Extending this results to all relations gives:

Proposition 5. *Every recursively enumerable relation is Σ_1 and every Σ_1 relation is recursively enumerable.*

A Recursive Pairing Function $\delta(x, y)$

A function f which assigns to each pair (x, y) of numbers a number $f(x, y)$ is said to be 1-1 (which is read as "one-to-one") if for every number z there is at most one pair (x, y) such that $f(x, y) = z$. In other words, if $f(x, y) = f(z, w)$, then $x = z$ and $y = w$. The function f is said to be *onto* the set N of numbers, if every number z is $f(x, y)$ for some pair (x, y). Thus f is 1-1 onto N iff every number z is $f(x, y)$ for one and only one pair (x, y).

We now want a *recursive* function $\delta(x, y)$ which is 1-1 onto N. [Note that such a function provides an enumeration of all ordered pairs of positive integers.] There are many ways to construct such a function. Here is one way: We start with $(1, 1)$ followed by the three pairs whose highest number is 2, in the order $(1, 2), (2, 2), (2, 1)$, followed by the five pairs whose highest number is 3, in the order $(1, 3), (2, 3), (3, 3), (3, 1), (3, 2)$, followed by $\ldots$ $(1, n), (2, n), \ldots, (n, n), (n, 1), (n, 2), \ldots, (n, n - 1)$, etc. We then take $\delta(x, y)$ to be the position that (x, y) takes in this ordering, which is the number of pairs which precede (x, y) plus 1. For example,

$$\delta(1, 1) = 1, \quad \delta(1, 2) = 2, \quad \delta(2, 2) = 3, \quad \delta(2, 1) = 4, \quad \delta(1, 3) = 5,$$

$$\delta(2, 3) = 6, \quad \delta(3, 3) = 7, \quad \delta(3, 1) = 8, \quad \delta(3, 2) = 9, \quad \delta(1, 4) = 10.$$

In fact, the following hold:
(a) If $m \leq n$, then $\delta(m, n) = (n - 1)^2 + m$.
(b) If $n < m$, then $\delta(m, n) = (n - 1)^2 + m + n$.

To see that (a) is true, we first note that for any n, the pairs that preceded $(1, n)$ in the ordering are those in which x and y are both less than n, and there are $(n - 1)^2$ such pairs [since there are $n - 1$ numbers (positive integers) that are less than n]. Thus $\delta(1, n) = (n - 1)^2 + 1$. Hence $\delta(2, n) = (n - 1)^2 + 2$, $\delta(3, n) = (n - 1)^2 + 3, \ldots, \delta(n, n) = (n - 1)^2 + n$. Thus, for any $m \leq n$, $\delta(m, n) = (n - 1)^2 + m$. This proves (a).

As for (b), since $\delta(m, m) = (m - 1)^2 + m$, then $\delta(m, 1) = (m - 1)^2 + m + 1$; $\delta(m, 2) = (m - 1)^2 + m + 2, \ldots, \delta(m, m - 1) = (m - 1)^2 + m + (m - 1)$. Thus for any $n \leq m$, $\delta(m, n) = (m - 1)^2 + m + n$.

Problem 7. Prove that the function $\delta(x, y)$ is recursive, in fact Σ_0.

The Inverse Functions K and L

Each number z is $\delta(x, y)$ for just one pair (x, y). We let $K(z)$ [to often be abbreviated Kz] be x, and let $L(z)$ [or Lz] be y. Thus $K\delta(x, y) = x$ and $L\delta(x, y) = y$.

Problem 8. (a) Prove that $\delta(Kz, Lz) = z$. (b) Prove that the functions $K(x)$ and $L(x)$ are Σ_0, and hence recursive.

The n-tupling Functions $\delta_n(x_1, \ldots, x_n)$

For each $n \geq 2$ we define the function $\delta_n(x_1, \ldots, x_n)$ by the following inductive scheme:

$$\delta_2(x_1, x_2) = \delta(x_1, x_2),$$
$$\delta_3(x_1, x_2, x_3) = \delta(\delta_2(x_1, x_2), x_3),$$
$$\vdots$$
$$\delta_{n+1}(x_1, \ldots, x_{n+1}) = \delta(\delta_n(x_1, \ldots, x_n), x_{n+1}).$$

An obvious induction argument shows that for each $n \geq 2$ the function $\delta_n(x_1, \ldots, x_n)$ is a 1-1 recursive function *onto* the set N of positive integers.

Problem 9. Prove the following:

(a) For any recursively enumerable set A and $n \geq 2$, the relation $\delta_n(x_1, \ldots, x_n) \in A$ is recursively enumerable.
(b) For any recursively enumerable relation $R(x_1, \ldots, x_n)$, the set A of all numbers $\delta_n(x_1, \ldots, x_n)$ such that $R(x_1, \ldots, x_n)$ is recursively enumerable.

The Inverse Functions K_i^n

For each $n \geq 2$ and $1 \leq i \leq n$, we define $K_i^n(x)$ as follows: $x = \delta_n(x_1, \ldots, x_n)$ for just one n-tuple $(x_1, \ldots, x_n)$, and we take $K_i^n(x)$ to be x_i. Thus $K_i^n(\delta_n(x_1, \ldots, x_n)) = x_i$.

Problem 10. Prove the following:

(a) $\delta_n(K_1^n(x), K_2^n(x), \ldots, K_n^n(x)) = x$.

(b) $K_1^2(x) = Kx$ and $K_2^2(x) = Lx$.

(c) If $x = \delta_{n+1}(x_1, \ldots, x_{n+1})$, then $Kx = \delta_n(x_1, \ldots, x_n)$ and $Lx = x_{n+1}$.

(d) If $i \leq n$ then $K_i^{n+1}(x) = K_i^n(Kx)$. If $i = n + 1$, then $K_i^{n+1}(x) = Lx$.

(e) For any recursively enumerable relation $R(x_1, \ldots, x_n, y_1, \ldots, y_m)$, there is a recursively enumerable relation $M(x, y)$ such that for all $x_1, \ldots, x_n, y_1, \ldots, y_m$, we have $R(x_1, \ldots, x_n, y_1, \ldots, y_m)$ iff $M(\delta_n(x_1, \ldots, x_n), \delta_m(y_1, \ldots, y_m))$.

III.   A Universal System

We now wish to construct a "universal" system (U), in which we can express all propositions of the form that such and such a number is in such and such a recursively enumerable number set, or that such and such an n-tuple of numbers stands in such and such a recursively enumerable relation on n-tuples of numbers. (We recall that we are using the word "number" to mean "positive integer".)

In preparation for the construction of the system (U), we need to "transcribe" all dyadic systems (elementary formal systems over the alphabet $\{1, 2\}$) into a single *finite* alphabet. We now define a *transcribed dyadic system* — to be abbreviated "TDS" — to be a system like a dyadic system, except that instead of taking our variables and predicates to be individual symbols, we take two signs, v and the "accent sign" $'$ and we define a *transcribed variable* — to be abbreviated "TS variable" — to be any string $v \alpha v$, where α is a string of accents. Similarly, we define a *transcribed predicate* — to be abbreviated "TS predicate" — to be a string of p's followed by a string of accents; the number of p's is to indicate the *degree* of the predicate. A transcribed dyadic system is then like a dyadic system, except that we use transcribed variables instead of single symbols for variables, and transcribed predicates instead of individual symbols for predicates. We thus have a single alphabet K_7, namely the symbols from which all TS variables and predicates, and eventually formulas, are constructed:

$$'\quad v\quad p\quad 1\quad 2\quad ,\quad \rightarrow$$

Note that the symbols 1 and 2, from which we will construct our dyadic numerals for the positive integers, correspond to the alphabet K in the definition of an elementary formal system given in *The Beginner's Guide* [Smullyan, 2014] and earlier here.

It is obvious that representability in an arbitrary dyadic system is equivalent to representability in a transcribed dyadic system.

We define "TS term", "atomic TS formula", "TS formula", "TS sentence" as we did "term", "atomic formula", "formula", "sentence", only using TS variables and TS predicates instead of variables and predicates, respectively.

We now construct the universal system (U) in the following manner: We first extend the alphabet K_7 to the alphabet K_8 by adding the symbol "$\wedge$". Then, by a *base* we mean a string of the form $X_1, \ldots, X_k$ (or a single formula X, if $k = 1$). Here X and $X_1, \ldots, X_k$ are all TS formulas. These TS formulas are called the *components* of the base $X_1 \wedge \ldots \wedge X_k$, or of the base X. Then we extend the alphabet K_8 to the alphabet K_9 by adding the symbol "$\vdash$", and we define a *sentence* of (U) to be an expression of the form $B \vdash X$, where B is a base and X is a TS sentence (a TS formula with no variables). We call the sentence $B \vdash X$ *true* if X is provable in that TDS whose axiom schemes are the components of B. Thus $X_1 \wedge \ldots \wedge X_k \vdash X$ is true iff X is provable in that TDS whose axiom schemes are $X_1, \ldots, X_k$.

By a *predicate H of* (U) [not to be confused with a TS predicate], we mean an expression of the form $B \vdash P$, where B is a base and P is a transcribed predicate. We define the *degree* of $B \vdash P$ to be the degree of P (i.e. the number of occurrences of p in P). A predicate H of the form $B \vdash P$ of (U) of degree n is said to *represent* the set of all n-tuples $(a_1, \ldots, a_n)$ of numbers such that the sentence $B \vdash Pa_1, \ldots, a_n$ is true. Thus a predicate $X_1 \wedge \ldots \wedge X_k \vdash P$ of degree n represents (in U) the set of all n-tuples $(a_1, \ldots, a_n)$ such that $Pa_1, \ldots, a_n$ is provable in that TDS whose axiom schemes are $X_1, \ldots, X_k$; thus it represents in (U) the same relation as P represents in the TDS whose axiom schemes are $X_1, \ldots, X_k$. Thus a relation or set is representable in (U) if and only if it is recursively enumerable. In this sense, (U) is called a *universal system* for all recursively enumerable relations.

We shall let T be the set of true sentences of (U), and let T_0 be the set of Gödel numbers of the members of T (using the *dyadic* Gödel numbering).

We now wish to show that the set T is formally representable in an elementary formal system over the alphabet K_9, and that the set T_0 is recursively enumerable. Our entire development of recursion theory is based on this crucial fact!

We are to construct an elementary formal system $\mathcal{W}$ over K_9 in which T is represented. To do this, we must first note that the implication sign of $\mathcal{W}$ must be different from the implication sign of transcribed systems. We will use "$\rightarrow$" for the implication sign of $\mathcal{W}$, and use "*imp*" for the implication sign of transcribed systems. Similarly, we will use the ordinary comma for the punctuation sign of $\mathcal{W}$ and use "*com*" for the punctuation sign of transcribed systems. For variables of $\mathcal{W}$, (not to be confused with transcribed variables), we will use the letters x, y, z, w, with or without subscripts. Predicates (not to be confused with predicates of (U) or transcribed predicates) are to be introduced as needed.

Problem 11. Construct $\mathcal{W}$ by successively adding predicates and axiom schemes to represent the following properties and relations:

$N(x)$: x is a number (dyadic numeral)

$Acc(x)$: x is a string of accents

$Var(x)$: x is a transcribed variable

$dv\,(xy)$: x and y are distinct TS variables

$P(x)$: x is a transcribed predicate

$t(x)$: x is a transcribed term

$F_0(x)$: x is a transcribed atomic formula

$F(x)$: x is a transcribed formula

$S_0(x)$: x is a transcribed atomic sentence

$Sub(x, y, z, w)$: x is a string compounded from TS variables, numerals, TS predicates, *com* and *imp*, and y is a TS variable, and w is the result of substituting z for every occurrence of y in x

$pt(x, y)$: x is a TS formula and y is the result of substituting numerals for some (but not necessarily all) TS variable of x (such a formula y is called a *partial instance* of x)

$in(x, y)$: x is a TS formula and y is the result of substituting numerals for *all* variables of x (such a y is an *instance* of x)

$B(x)$: x is a base

$C(x, y)$: x is a component of the base y

$T(x)$: x is a true sentence of (U)

Since T_0 is formally representable, then T_0 is Σ_1, hence recursively enumerable.

The Recursive Unsolvability of (U)

We now wish to show that the complement $\overline{T_0}$ of T_0 is not recursively enumerable, and thus that the set T_0 is recursively enumerable, but not recursive.

For any set A of numbers, define a sentence X of (U) to be a *Gödel sentence for A* if either X is true and its Gödel number X_0 is in A, or X is false and its Gödel number is not in A. In other words, a sentence X of (U) is a *Gödel sentence for A* if X is true iff $X_0 \in A$. There obviously cannot be a Gödel sentence for $\overline{T_0}$, for such a sentence would be true iff it were not true (i.e. iff its Gödel number was not in T_0), which is impossible. Thus, to show that $\overline{T_0}$ is not recursively enumerable, it suffices to show that for any recursively enumerable set, there is a Gödel sentence.

To this end, for any string X of symbols of K_7, we define its *norm* to be XX_0, where X_0 is the Gödel number of X. Any dyadic numeral n itself has a Gödel number n_0, hence a norm nn_0. We recall that our dyadic Gödel numbering has the advantage of being an isomorphism with respect to concatenation, i.e. if n is the Gödel number of X and m is the Gödel number of Y, then nm is the Gödel number of XY. Therefore, if n is the Gödel number of X, then nn_0 is the Gödel number of Xn (the norm of X).

Thus the Gödel number of the norm of X is the norm of the Gödel number of X.

For any number x, we let *normx* be the norm of x.

Problem 12. Show that the relation *normx* $= y$ is recursively enumerable.

For any set A of numbers we let $A^{\#}$ be the set of all numbers whose norm is in A.

Lemma. *If A is recursively enumerable, so is $A^{\#}$.*

Problem 13. Prove the above lemma.

Problem 14. Using the above lemma, now prove that if A is recursively enumerable, then there is a Gödel sentence for A.

As we have seen, there cannot be a Gödel sentence for $\overline{T_0}$, and thus by Problem 14, the set $\overline{T_0}$ cannot be recursively enumerable. And so the set T_0 is recursively enumerable, but not recursive.

Solutions to the Problems of Chapter 4

1. (a) We have already proved this for when R_1 and R_2 are sets (i.e. relations of degree 1), but the proof for relations of degree n is essentially the same: Suppose R_1 and R_2 are relations of degree n and that both are formally representable over K. There they are representable in a common elementary formal system over K. In this system let P_1 represent R_1 and let P_2 represent R_2. Then take a new predicate Q and add the following two axioms:

$$P_1 x_1, \ldots, x_n \to Q x_1, \ldots, x_n$$
$$P_2 x_1, \ldots, x_n \to Q x_1, \ldots, x_n.$$

Then Q represents $R_1 \cup R_2$.

To represent $R_1 \cap R_2$, instead of the above two axioms, add the following single axiom:

$$P_1 x_1, \ldots, x_n \to P_2 x_1, \ldots, x_n \to Q x_1, \ldots, x_n.$$

Then Q will represent $R_1 \cap R_2$.

(b) Let P represent the relation $R(x_1, \ldots, x_n, y)$ in (E). Take a new predicate Q of degree n and add the following axiom scheme:

$$Px_1, \ldots, x_n, y \to Qx_1, \ldots, x_n.$$

Then Q represents the relation

$$\exists y R(x_1, \ldots, x_n, y) \text{ in } (E).$$

(c) Let P represent the relation $R(x_1, \ldots, x_k)$ in (E). Take a new predicate Q and add the following axiom scheme:

$$P\alpha_1, \ldots, \alpha_k \to Qx_1, \ldots, x_n.$$

Then Q represents the relation

$$\lambda x_1, \ldots, x_n R(\alpha_1, \ldots, \alpha_k) \text{ in } (E).$$

2. Complementation is trivial. As for unions and intersections, suppose that W_1 and W_2 are of the same degree and are both solvable over K. Then W_1, W_2, $\overline{W_1}$ and $\overline{W_2}$ are all formally representable over K. We already know from Problem 1 that $W_1 \cup W_2$ and $W_1 \cap W_2$ are formally representable over K. It remains to show that the complements of $W_1 \cup W_2$ and $W_1 \cap W_2$ are formally representable over K. But these complements are $\overline{W_1 \cup W_2}$ and $\overline{W_1 \cap W_2}$, which we can use our knowledge of Boolean sets to see to be equal to $\overline{W_1} \cap \overline{W_2}$ and $\overline{W_1} \cup \overline{W_2}$, respectively. Since these latter sets are the intersections and unions of sets we know to be formally representable over K, we now see that $\overline{W_1 \cup W_2}$ and $\overline{W_1 \cap W_2}$ are formally representable over K, and thus that the union or intersection of two sets that are solvable over K is solvable over K.

Now assume that the $R(x_1, \ldots, x_k)$ is solvable over K. Then its complement, which we will denote by $\overline{R}(x_1, \ldots, x_k)$ is also solvable over K, since the two are complements of each other. We have seen that the explicit transformations $\lambda x_1, \ldots, x_n R(\alpha_1, \ldots, \alpha_k)$ and $\lambda x_1, \ldots, x_n \overline{R}(\alpha_1, \ldots, \alpha_k)$, where each α_i is one of the variables $x_1, \ldots, x_n$ or a constant, are also formally representable over K. But each of these is the complement of the other, so we see that the collection of sets and relations solvable over K is closed under explicit transformation as well.

3. (a) Suppose that the relation $R(x_1, \ldots, x_n, y)$ is recursively enumerable. Let D be a dyadic system in which R is represented by the

predicate P, $x < y$ is represented by E, and the relation "the successor of x is y" is represented by S.

Let $W(x_1, \ldots, x_n, y)$ be the relation $(\forall z \leq y)R(x_1, \ldots, x_n, z)$. To represent the relation W, take a new predicate Q and add the following two axiom schemes:

$$Px_1, \ldots, x_n, 1 \to Qx_1, \ldots, x_n, 1$$

$$Qx_1, \ldots, x_n, z \to Sz, z' \to Ez, y \to Qx_1, \ldots, x_n, z'.$$

Then Q represents W.

As for the relation $(\exists z \leq y)R(x_1, \ldots, x_n, z)$ (as a relation between $x_1, \ldots, x_n, y$), it is equivalent to the relation $\exists z(z \leq y \wedge R(x_1, \ldots, x_n, z)$. Let $S(x_1, \ldots, x_n, y, z)$ be the relation $z \leq y \wedge R(x_1, \ldots, x_n, z)$. It must be recursively enumerable, because it is the intersection $S_1 \cap S_2$ of the relations S_1 and S_2 where

$$S_1 = \lambda x_1, \ldots, x_n, y, z\,(z \leq y)$$

and

$$S_2 = \lambda x_1, \ldots, x_n, y, z\, R(x_1, \ldots, x_n, z).$$

Since S_1 and S_2 are respectively definable from the recursively enumerable relations $z \leq y$ and $R(x_1, \ldots, x_n, z)$ by explicit transformation, they are recursively enumerable, and so is their intersection S. Thus so is the existential quantification $\exists z S(x_1, \ldots, x_n, z)$, which is the relation $(\exists z \leq y)R(x_1, \ldots, x_n, z)$.

(b) Suppose R is recursive. Thus R and its complement $\overline{R}$ are both recursively enumerable. Since R is recursively enumerable, so are $\forall_F(R)$ and $\exists_F(R)$ [by (a)]. Since $\overline{R}$ is recursively enumerable, so are $\forall_F(\overline{R})$ and $\exists_F(\overline{R})$ [again by (a)]. But as the reader can verify, $\forall_F(\overline{R})$ and $\exists_F(\overline{R})$ are respectively the complements of $\exists_F(R)$ and $\forall_F(R)$, and so the complements of $\exists_F(R)$ and $\forall_F(R)$ are both recursively enumerable, which means that $\exists_F(R)$ and $\forall_F(R)$ are both recursive.

4. Let us first note that for any relation $R(x_1, \ldots, x_n, y, z)$ the following conditions are equivalent:

(1) $\exists y \exists z R(x_1, \ldots, x_n, y, z)$

(2) $\exists w(\exists y \leq w)(\exists z \leq w)R(x_1, \ldots, x_n, y, z)$

It is obvious that (2) implies (1). Now, suppose that (1) holds. Then there are numbers y and z such that $R(x_1, \ldots, x_n, y, z)$. Let

w be the maximum of y and z. Then $y \leq w$ and $z \leq w$, and so $(\exists y \leq w)(\exists z \leq w)R(x_1, \ldots, x_n, y, z)$ holds for such a number w. Hence (2) holds.

Now suppose $R(x_1, \ldots, x_n, y)$ is Σ_1. Then there is a Σ_0 relation $S(x_1, \ldots, x_n, y, z)$ such that $R(x_1, \ldots, x_n, y)$ is equivalent to $\exists z S(x_1, \ldots, x_n, y, z)$. Thus $\exists y R(x_1, \ldots, x_n, y)$ is equivalent to $\exists y \exists z S(x_1, \ldots, x_n, y, z)$, which, as shown above, is Σ_1 [since the relation $(\exists y \leq w)(\exists z \leq w)S(x_1, \ldots, x_n, y, z)$ is Σ_0].

5. $x \in A'$ iff $f(x) \in A$, which is true iff $\exists y(f(x) = y \wedge y \in A)$. We first show that the relation $f(x) = y \wedge y \in A$ is Σ_1.

Well, since the relation $f(x) = y$ is Σ_1, then there is a Σ_0 relation $S(x, y, z)$ such that $f(x) = y$ iff $\exists z S(x, y, z)$.

Since the set A is Σ_1, there is a Σ_0 relation $R(x, y)$ such that $x \in A$ iff $\exists w R(x, w)$. Thus $f(x) = y \wedge y \in A$, iff

$$\exists z S(x, y, z) \wedge \exists w R(x, w),$$

which is true iff $\exists z \exists w(S(x, y, z) \wedge R(x, w))$. Since the condition $S(x, y, z) \wedge R(x, w)$ is Σ_0, then $\exists w(S(x, y, z)R(x, w))$, is Σ_1, hence so is the condition $\exists z \exists w(S(x, y, z)R(x, w))$ [by Problem 4].

Thus the condition $f(x) = y \wedge y \in A$ is Σ_1, and therefore so is $\exists y(f(x) = y \wedge y \in A)$ [again by Problem 4]. Thus A' is Σ_1.

6. Our goal is to prove that the relation $g(x) = y$ is Σ_1. First let $R_1(x, y)$ be the Σ_0 relation $(x = 1 \wedge y = 12) \vee (x = 2 \wedge y = 122)$ Thus $R_1(x, y)$ holds iff x is the single digit 1 or the single digit 2 and $g(x) = y$.

Let $R_2(x, y, x_1, y_1)$ be the Σ_0 relation

$$(x = x_1 1 \wedge y = y_1 12) \vee (x = x_1 2 \wedge y = y_1 122). \quad .$$

Call a set S of ordered pairs of numbers *special*, if for every pair (x, y) in S, either $R_1(x, y)$ or $(\exists x_1 \leq x)(\exists y_1 \leq y)((x_1, y_1) \in S \wedge R_2(x, y, x_1, y_1))$. It is easily seen by mathematical induction on the length of x (i.e. the number of occurrences of 1 or 2) that if S is special, then for every pair (x, y) in S, it must be that $g(x) = y$.

Next we must show that conversely, if $g(x) = y$, then (x, y) belongs to some special set.

Well, the x in $g(x) = y$ must be of the form $x = d_1 d_2 \ldots d_n$, where each d_i is the digit 1 or the digit 2. For each $i \leq n$, we let g_i be the dyadic Gödel number of d_i (thus, if $d_i = 1$, then $g_i = 12$ and if $d_i = 2$, then $g_i = 122$). Now let S_x be the set

$$\{(d_1, g_1), (d_1 d_2, g_1 g_2), \ldots, (d_1 d_2 \ldots d_n, g_1 g_2 \ldots g_n)\}.$$

The set S_x is obviously special, and contains the pair $(x, g(x))$ [which is the pair $(d_1 d_2 \ldots d_n, g_1 g_2 \ldots g_n)$]. Thus if $g(x) = y$, then $(x, y) \in S_x$ and so (x, y) belongs to a special set. Thus we have:

(1) $g(x) = y$ iff (x, y) belongs to a special set.

We now use Lemma K. For any number z, we let S_z be the set of all pairs (x, y) such that $K(x, y, z)$ holds. Thus $(x, y) \in S_z$ iff $K(x, y, z)$. Since $(x, y) \in S_z$ iff $K(x, y, z)$, then to say that S_z is special is to say that for all x and y, $K(x, y, z)$ implies that the following Σ_0 condition — call it $A(x, y, z)$ — holds:

$$R_1(x, y) \vee (\exists x_1 \leq x)(\exists y_1 \leq y)(K(x_1, y_1, z) \wedge R_2(x, y, x_1, y_1)).$$

Thus S_z is special iff the following condition holds:

$$C(z)\colon \forall x \forall y (K(x, y, z) \supset A(x, y, z)).$$

However, $C(z)$ is equivalent to the following Σ_0 condition:

$$P(z)\colon (\forall x \leq z)(\forall y \leq z)(K(x, y, z) \supset A(x, y, z)).$$

To see the equivalence, it is obvious that $C(z)$ implies $P(z)$. To see the converse, suppose $P(z)$ holds. We must show that $C(z)$ holds. Let x and y be arbitrary numbers. We need to show that if $K(x, y, z)$ holds, then so does $A(x, y, z)$. If $K(x, y, z)$ holds, then it must also be true that $x \leq z$ and $y \leq z$ [by the second condition in Lemma K], and therefore $A(x, y, z)$ holds [by the assumption $P(z)$]. Thus, for all x and y, $K(x, y, z) \supset A(x, y, z)$, and so $C(z)$ holds.

This proves that $C(z)$ is equivalent to $P(z)$, and hence S_z is special iff $P(z)$ holds.

Finally, $g(x) = y$ iff (x, y) belongs to some special set, which is true iff

$$\exists z((x, y) \in S_z \wedge S_z \text{ is special}),$$

which is true iff

$$\exists z(K(x, y, z) \wedge P(z)).$$

Thus the relation $g(x) = y$ is Σ_1 (since $K(x, y, z) \wedge P(z)$ is Σ_0), which completes the proof that every recursively enumerable set A is Σ_1.

7. $\delta(x, y) = z$ iff

$$(x \leq y \wedge z = (y-1)^2 + x) \vee (y < x \wedge z = (x-1)^2 + x + y).$$

The condition $z = (y-1)^2 + x$ can be written

$$(\exists w \leq z)(y = w + 1 \wedge z = w^2 + x).$$

The condition $z = (x-1)^2 + x + y$ can be written

$$(\exists w \leq z)(x = w + 1 \wedge z = w^2 + x + y).$$

As the formula expressing $\delta(x, y) = z$ can be rewritten as a Σ_0 formula, the function δ is Σ_0, and consequently recursive [by Proposition 2].

8. (a) $z = \delta(x, y)$ for some x and y. Then $Kz = x$ and $Lz = y$, so that $\delta(x, y) = \delta(Kz, Lz)$. Thus z [which is $\delta(x, y)$] equals $\delta(Kz, Lz)$.

   (b) $K(z) = x$ iff $(\exists y \leq z)(\delta(x, y) = z)$ and $L(z) = y$ iff $(\exists x \leq z)(\delta(x, y) = z)$. (It is not difficult to verify that if $\delta(x, y) = z$, both $x \leq z$ and $y \leq z$.)

9. Conceptually, we know the following:

   (a) $\delta_n(x_1, \ldots, x_n) \in A$ iff $\exists y(\delta_n(x_1, \ldots, x_n) = y \wedge y \in A)$.

   (b) $x \in A$ iff $\exists x_1 \ldots \exists x_n(x = \delta_n(x_1, \ldots, x_n) \wedge R(x_1, \ldots, x_n))$.

To see that this understanding solves the problem, we can amplify as follows:

(a) When we say the set A is recursively enumerable, what we mean is that there's a relation of degree 1, $R(x)$, in the form of a Σ_1 formula F_1 with one free variable x (and one existential quantifier) such that the set of numbers in A is the set of all numbers x for which $F_1(x)$ holds. Also, since $\delta_n(x_1, \ldots, x_n) = y$ has been shown to be recursive, there is a Σ_0 formula F_2 containing the free variables $x_1, \ldots, x_n, y$ which holds for an $n + 1$-tuple of numbers iff the equation $\delta_n(x_1, \ldots, x_n) = y$ is true (when the $n + 1$ numbers are "plugged in" appropriately). So $\delta_n(x_1, \ldots, x_n) \in A$ can be expressed by the formula

$$\exists y(F_2(x_1, \ldots, x_n, y) \wedge F_1(y)).$$

This can be seen to be a Σ_1 formula, for the single unbound quantifier in F_1 can be brought to the front in a way the reader should be familiar with by now.

(b) Since the relation $R(x_1, \ldots, x_n)$ is recursively enumerable, there is a Σ_1 formula F_1 with the free variables $x_1, \ldots, x_n$ such that $F_1(x_1, \ldots, x_n)$ holds iff $R(x_1, \ldots, x_n)$ is true. Assuming F_2 as in part (a) of this problem, $x \in A$ can be expressed by the formula $\exists x_1 \ldots \exists x_n (F_2(x_1, \ldots, x_n, x) \wedge F_1(x_1, \ldots, x_n))$.

10. (a) For a given x, since δ_n is 1-1 and onto, x is equal to $\delta_n(x_1, \ldots, x_n)$ for just one n-tuple $x_1, \ldots, x_n$. Moreover for each $i \leq n$, $K_i^n(x)$ is defined to be x_i. So

$$\delta_n(K_1^n(x), K_2^n(x), \ldots, K_n^n(x)) = \delta_n(x_1, \ldots, x_n) = x.$$

(b) Since $x = \delta_2(Kx, Lx)$ [by the definition of δ_2 and Problem 8], we have that $K_1^2(x) = K_1^2(\delta_2(Kx, Lx)) = Kx$ and $K_2^2(\delta_2(Kx, Lx)) = Lx$.

(c) Suppose $x = \delta_{n+1}(x_1, \ldots, x_{n+1})$. Then [by the definition of δ_{n+1}], $x = \delta(\delta_n(x_1, \ldots, x_n), x_{n+1})$. Thus $Kx = \delta_n(x_1, \ldots, x_n)$ and $Lx = x_{n+1}$.

(d) Let $x = \delta_{n+1}(x_1, \ldots, x_n, x_{n+1})$ for some $x_1, \ldots, x_{n+1}$. Then by definition

$$K_i^{n+1}(x) = x_i.$$

Now suppose $i \leq n$. Then $Kx = \delta_n(x_1, \ldots, x_n)$ [by (c)], so that

$$K_i^n(Kx) = K_i^n(\delta_n(x_1, \ldots, x_n)) = x_i.$$

But also $K_i^{n+1}(x) = x_i$. So $K_i^{n+1}(x) = K_i^n(Kx)$ [for $i \leq n$]. Now, $K_{n+1}^{n+1}(x) = x_{n+1} = Lx$ [by (c)].

(e) Take $M(x, y) = R(K_1^n(x), \ldots, K_n^n(x)K_1^m(y), \ldots, K_m^m(y))$. Then $R(x_1, \ldots, x_n, y_1, \ldots, y_m)$ iff $M(\delta_n(x_1, \ldots, x_n), \delta_m(y_1, \ldots, y_m))$.

11. $N(x)$: x is a number

$$N\,1$$
$$N\,2$$
$$N\,x \to N\,y \to N\,y$$

$Acc(x)$: x is a string of accents

$$Acc\,'$$
$$Acc\,x \to Acc\,x'$$

$Var(x)$: x is a TS variable

$$Acc\,x \to Var\,vx$$

$dv(x, y)$: x and y are distinct TS variable

$$Acc\,x \rightarrow Acc\,y \rightarrow dv\,vxv, vxyv$$
$$dv\,x, y \rightarrow dv\,y, x$$

$P(x)$: x is a TS predicate

$$Acc\,x \rightarrow P\,px$$
$$P\,x \rightarrow P\,px$$

$t(x)$: x is a TS term

$$N\,x \rightarrow t\,x$$
$$Var\,x \rightarrow t\,x$$
$$t\,x \rightarrow t\,y \rightarrow t\,xy$$

$F_0(x)$: x is a TS atomic formula

$$Acc\,x \rightarrow t\,y \rightarrow F_0\,pxy$$
$$F_0\,x \rightarrow t\,y \rightarrow F_0\,pxcomy$$

$F(x)$: x is a TS formula

$$F_0\,x \rightarrow F\,x$$
$$F_0\,x \rightarrow F\,y \rightarrow F\,ximpy$$

$S_0(x)$: x is a TS atomic sentence

$$Acc\,x \rightarrow N\,y \rightarrow S_0\,pxy$$
$$S_0\,x \rightarrow N\,y \rightarrow S_0\,pxcomy$$

$S(x)$: x is a TS sentence

$$S_0\,x \rightarrow S\,y$$
$$S_0\,x \rightarrow S\,y \rightarrow S\,ximpy$$

$Sub(x, y, z)$: x is a string, y a TS variable, z a numeral, and...

$$N\,x \rightarrow Var\,y \rightarrow N\,z \rightarrow Sub\,x, y, z, x$$
$$Var\,x \rightarrow N\,z \rightarrow Sub\,x, x, z, z$$
$$dv\,x, y \rightarrow N\,z \rightarrow Sub\,x, y, z, x$$
$$P\,x \rightarrow Var\,y \rightarrow N\,z \rightarrow Sub\,x, y, z, x$$
$$Var\,y \rightarrow N\,z \rightarrow Sub\,com, y, z, com$$
$$Var\,y \rightarrow N\,z \rightarrow Sub\,imp, y, z, imp$$
$$Sub\,x, y, z, w \rightarrow Sub\,x_1, y, z, w_1 \rightarrow Sub\,xx_1, y, z, ww_1$$

$pt(x, y)$: x is a TS atomic formula and y is a partial instance of x

$$F\,x \rightarrow Sub\,x, y, z, w \rightarrow pt\,x, w$$
$$pt\,x, y \rightarrow pt\,y, z \rightarrow pt\,x, z$$

$in(x, y)$: x is a TS atomic formula and y is an instance of x

$$pt\, x, y \to S\, y \to in\, x, y$$

$B(x)$: x is a TS base

$$F\, x \to B\, x$$
$$B\, x \to F\, x \to B\, x \wedge y$$

$C(x, y)$: x is a component of the base and y

$$F\, x \to C\, x, x$$
$$C\, x, y \to F\, z \to C\, x, y \wedge z$$
$$C\, x, y \to F\, z \to C\, x, z \wedge y$$

$T(x)$: x is a true TS sentence

$$C\, x, y \to in\, x, x_1 \to T\, y \vdash x_1$$
$$T\, y \vdash x \to T\, y \vdash x\, imp\, z \to S_0 x \to T\, y \vdash z$$

12. Let D be a dyadic system with predicate G and the following axiom schemes:

$$G\, 1, 12$$
$$G\, 2, 122$$
$$G\, x \to G\, y \to G\, xy$$

Then G represents the relation "the Gödel number of x is y". Now add a predicate M and the axiom scheme:

$$G\, x, y \to M\, x, xy.$$

Then M represents the relation "$norm(x) = y$".

13. Suppose A is recursively enumerable. Let D be a dyadic system in which "A" represents the set A and "M" represents the relation "the norm of x is y". Add a predicate "B" and the axiom scheme

$$M\, x, y \to A\, y \to B\, x.$$

Then B represents the set $A^{\#}$.

Remark. Without using elementary formal systems or dyadic system, one can directly show that if A is Σ_1, so is $A^{\#}$, and hence that if A is recursively enumerable, so is $A^{\#}$. As a good exercise, show that if A is Σ_1, so is $A^{\#}$.

14. The following beautiful proof is a modification of Gödel's famous diagonal argument in his incompleteness theorem proof, and does for recursion theory what Gödel did for mathematical systems.

Suppose A is recursively enumerable. Then so is $A^\#$ (by the Lemma). Let H be a predicate of (U) that represents $A^\#$. Then for every number n,

$$Hn \text{ is true iff } n \in A^\#, \text{ which if true iff } nn_0 \in A.$$

Thus Hn is true iff $nn_0 \in A$.

We take for n the Gödel number h of H, and obtain that Hh is true iff $hh_0 \in A$.

However, hh_0 is the Gödel number of Hh! Thus Hh is true iff its Gödel number is in A, and if thus a Gödel sentence for A.

Chapter 5

Some Recursion Theory

I. Enumeration and Iteration Theorems

We now turn to two basic theorems of recursion theory — the enumeration theorem, due to Emil Post [1943; 1944] and to Stephen Kleene [1952], and the iteration theorem, due to Kleene [1952] and independently discovered by Martin Davis [1958, 1982]. Once we have established these two results, the universal system (U) of the last chapter will have served its purpose, and will no longer be needed.

The Enumeration Theorem

We wish to arrange all recursively enumerable sets in an infinite sequence $\omega_1, \omega_2, \ldots, \omega_n, \ldots$ (allowing repetitions) in such a way that the relation "ω_x contains y" is a recursively enumerable relation between x and y.

We recall that for any number (dyadic numeral) y, we are letting y_0 denote the dyadic Gödel number of y. For any x and y we now define $r(x, y)$ to be xy_0 [x followed by y_0].

We use the letter T to denote the set of true sentences of the universal system (U) and use T_0 to denote the set of Gödel numbers of the sentences in T. We now define ω_n to be the set of all x such that $r(n, x) \in T_0$. Thus $x \in \omega_n$ iff $r(n, x) \in T_0$.

Problem 1. (a) Prove that ω_n is recursively enumerable. (b) Prove that for every recursively enumerable set A, there is some n such that $A = \omega_n$.

We shall call n an *index* of a set A if $A = \omega_n$. We have just shown that a set A is recursively enumerable if and only if it has an index, and so we have succeeded in arranging all recursively enumerable sets in an infinite sequence $\omega_1, \omega_2, \ldots, \omega_n, \ldots$ in such a manner that the relation ω_x contains y (which is the relation $r(x, y) \in T_0$) is recursively enumerable.

We next wish to show that for each $n \geq 2$, we can enumerate all recursively enumerable relations of degree n in an infinite sequence $R_1^n, R_2^n, \ldots, R_i^n, \ldots$ such that the relation $R_x^n(y_1, \ldots y_n)$ is a recursively enumerable relation among the variables $x, y_1, \ldots, y_n$. For this, it will be useful to use the indexing of recursively enumerable sets that we already have, along with the pairing function $\delta(x, y)$ and the n-tupling functions $\delta_n(x_1, \ldots, x_n)$. We simply define $R_i^n(x_1, \ldots, x_n)$ to be $\delta_n(x_1, \ldots, x_n) \in \omega_i$. Since ω_i is recursively enumerable and the function $\delta_n(x_1, \ldots, x_n)$ is also recursively enumerable, so is $R_i^n(x_1, \ldots, x_n)$ [by (a) of Problem 9 of Chapter 4]. Also, for any recursively enumerable relation $R(x_1, \ldots, x_n)$, the set A of numbers $\delta_n(x_1, \ldots, x_n)$ such that $R(x_1, \ldots, x_n)$ holds is recursively enumerable [by (b) of Problem 9 of Chapter 4]. Thus the set A has some index i so that $R(x_1, \ldots, x_n)$ holds iff $\delta_n(x_1, \ldots, x_n) \in \omega_i$, which is true iff $R_i^n(x_1, \ldots, x_n)$. Thus $R = R_i^n$. Thus every recursively enumerable relation has an index. [We also write R_n^1 for $x \in \omega_n$, since we are regarding number sets as relations of degree 1.]

We have thus proved:

Theorem E. [The Enumeration Theorem] *For each n, the set of all recursively enumerable relations of degree n can be enumerated in an infinite sequence $R_1^n, R_2^n, \ldots, R_x^n, \ldots$ such that the relation $R_x^n(y_1, \ldots y_n)$, as a relation among $x, y_1, \ldots, y_n$, is recursively enumerable.*

For each n we let $U^{n+1}(x, y_1, \ldots y_n)$ be the relation $R_x^n(y_1, \ldots, y_n)$ [as a relation among $x, y_1, \ldots, y_n$]. These relations are referred to as *universal relations*: U^2 is the universal relation for all recursively enumerable sets, U^3 is the universal relation for all recursively enumerable relations of degree 2, and so forth.

We sometimes write $R_x(y_1, \ldots, y_n)$ instead of $R_x^n(y_1, \ldots, y_n)$, since the number of arguments makes the superscript n clear. And we write $U(x, y_1, \ldots, y_n)$ instead of $U^{n+1}(x, y_1, \ldots, y_n)$.

Iteration Theorems

The second basic tool of recursion theory is the iteration theorem, which we will shortly state and prove in both a simple and more general form. To fully appreciate the significance and power of these theorems, the reader should first try solving the following exercises:

Exercise 1. We know that for any two recursively enumerable sets A and B, the union $A \cup B$ is recursively enumerable. Moreover, the process of finding this union is *effective*, in the sense that there is a recursive function $\varphi(i,j)$ such that if i is an index of A and j is an index of B, then $\varphi(i,j)$ is an index of $A \cup B$ — in other words there is a recursive function $\varphi(i,j)$ such that for all numbers i and j, $\omega_{\varphi(i,j)} = \omega_i \cup \omega_j$. Prove this.

For any relation $R(x,y)$, by its *inverse* is meant the relation $\overset{\vee}{R}$ satisfying the condition $\overset{\vee}{R}(x,y)$ iff $R(x,y)$.

Exercise 2. Show that there is a recursive function $t(i)$ such that for every number i the relation $R_{t(i)}(x,y)$ is the inverse of the relation $R_i(x,y)$, i.e. $R_{t(i)}(x,y)$ iff $R_i(y,x)$. [This can be paraphrased: Given an index of a recursively enumerable relation of two arguments, one can effectively find an index of its inverse.]

For any function $f(x)$ and number set A, by $f^{-1}(A)$ is meant the set of all numbers n such that $f(n) \in A$. Thus $n \in f^{-1}(A)$ iff $f(n) \in A$.

Exercise 3. Prove that for any recursive function $f(x)$, there is a recursive function $\varphi(i)$ such that for every number i, it is the case that

$$\omega_{\varphi(i)} = f^{-1}(\omega_i).$$

The reader who has solved the above exercises has probably had to go back to the universal system (U) each time. The iteration theorem, to which we now turn, will free us from this once and for all!

Theorem 1. [Simple Iteration Theorem] *For any recursively enumerable relation $R(x,y)$ there is a recursive function $t(i)$ such that for all numbers x and i, $x \in \omega_{t(i)}$ iff $R(x,i)$.*

Problem 2. Prove Theorem 1. [**Hint:** The recursively enumerable relation $R(x, y)$ is represented in the universal system (U) by some predicate H. Consider the Gödel number h of H, as well as the Gödel number c of the punctuation symbol *com*.]

Theorem 2. [Iteration Theorem] *For any recursively enumerable relation $R(x_1, \ldots, x_m, y_1, \ldots, y_n)$ there is a recursive function $\varphi(i_1, \ldots, i_n)$ such that for all $x_1, \ldots, x_m, i_1, \ldots, i_n$,*

$$R(x_1, \ldots, x_m, i_1, \ldots, i_n) \text{ iff } R_{\varphi(i_1,\ldots,i_n)}(x_1, \ldots, x_m).$$

One way to prove Theorem 2 is to use Theorem 1 and the fact that for any recursively enumerable relation $R(x_1, \ldots, x_m, y_1, \ldots, y_n)$ there is a recursively enumerable relation $M(x, y)$ such that $R(x_1, \ldots, x_m, y_1, \ldots, y_n)$ iff $M(\delta_m(x_1, \ldots, x_m), \delta_n(y_1, \ldots, y_n))$ [by Problem 10 (e) of Chapter 4].

Problem 3. Prove Theorem 2.

Let us now see how the iteration theorem provides simple and elegant solutions to the three recently given exercises.

In Exercise 1, let $R(x, y_1, y_2)$ be the recursively enumerable relation

$$x \in \omega_{y_1} \vee x \in \omega_{y_2}.$$

Then, by the iteration theorem, there is a recursive function $\varphi(i_1, i_2)$ such that for all numbers x, i_1, i_2,

$$x \in \omega_{\varphi(i_1,i_2)} \text{ iff } R(x, i_1, i_2), \text{ which is true iff } x \in \omega_{i_1} \vee x \in \omega_{i_2}.$$

Therefore, $\omega_{\varphi(i_1,i_2)} = \omega_{i_1} \cup \omega_{i_2}$.

For Exercise 2, let $R(x_1, x_2, y)$ be the recursively enumerable relation $R_y(x_2, x_1)$. It is recursively enumerable because it is explicitly defined from the recursively enumerable relation $U(x_1, x_2, y)$. Then by the simple iteration theorem there is a recursive function $\varphi(i)$ such that for all x and i, $R_{\varphi(i)}(x_1, x_2)$ iff $R(x_1, x_2, i)$, which is true iff $R_i(x_2, x_1)$.

For Exercise 3, given a recursive function $f(x)$, consider the recursively enumerable relation $R(x, y)$ – viz. $f(x) \in \omega_y$ [it is recursively enumerable since it is equivalent to $\exists z(f(x) = z \wedge z \in \omega_y)$]. By the simple iteration theorem there is a recursive function $t(i)$ such that for all x and i, $x \in \omega_{t(i)}$ iff $R(x, i)$ which is true iff $f(x) \in \omega_i$. Thus $\omega_{t(i)} = f^{-1}(\omega_i)$.

More serious applications of the iteration theorem will soon follow.

Maximal Indexing

A denumerable sequence $A_1, A_2, \ldots, A_n, \ldots$ of all recursively enumerable sets is called a *recursively enumerable indexing* if the relation $x \in A_y$ is recursively enumerable. It is called a *maximal* indexing if for every recursively enumerable indexing $B_1, B_2, \ldots, B_n, \ldots$ there is a recursive function $f(i)$ such that for all i the condition $A_{f(i)} = B_i$ holds. [Informally, this means that given a B-index of a recursively enumerable set, we can effectively find an A-index of the set.]

Another application of the iteration theorem yields:

Theorem 3. *The enumeration $\omega_1, \omega_2, \ldots, \omega_n, \ldots$ given by the enumeration theorem in maximal.*

Problem 4. Prove Theorem 3.

The following theorem is easily obtained from the iteration theorem, combined with the enumeration theorem.

Theorem 4. [Uniform Iteration Theorem] *For any numbers m and n, there is a recursive function $\varphi(i, i_1, \ldots, i_n)$ [sometimes written $S_n^m(i, i_1, \ldots, i_n)$] such that for all $x_1, \ldots, x_m, i_1, \ldots, i_n$,*

$$R_{\varphi(i, i_1, \ldots i_n)}(x_1, \ldots, x_m) \text{ iff } R_i(x_1, \ldots, x_m, i_1, \ldots, i_n).$$

Problem 5. Prove Theorem 4.

II. Recursion Theorems

Recursion theorems, which can be stated in many forms, and are sometimes called *fixed point theorems*, have many applications to recursion theory and metamathematics. To illustrate their rather startling nature, consider the following mathematical "believe it or nots."

Which of the following propositions, if true, would surprise you?

1. There is a number n such that $\omega_n = \omega_{n+1}$.
2. For any recursive function $f(x)$, there is a number n such that $\omega_{f(n)} = \omega_n$.
3. There is a number n such that ω_n contains n as its only member, i.e. $\omega_n = \{n\}$.

4. For any recursive function $f(x)$ there is a number n such that $\omega_n = \{f(n)\}$.

5. There is a number n such that for every number x, $x \in \omega_n$ iff $n \in \omega_x$.

Well, believe it or not, all the above statements are true! They are all special cases of the theorem that follows:

Consider a recursively enumerable relation $R(x, y)$. Now consider a number n and the set $x: R(x, n)$, i.e. the set of all numbers x such that $R(x, n)$ holds. This set is recursively enumerable, hence is ω_{n^*} for some n^*. Given the relation R, what n^* is depends on what n is. Wouldn't it be curious if n could be chosen such that n^* is n itself? Well, this is indeed so:

Theorem 5. [Weak Recursion Theorem] *For any recursively enumerable relation* $R(x, y)$, *there is a number* n *such that* $\omega_n = \{x : R(x, n)\}$ [*in other words, for all* x, *we have* $n \in \omega_n$ *iff* $R(x, n)$].

I will give the beginning of the proof, and leave it as a problem for the reader to finish it.

The relation $R_y(x, y)$, as a relation between x and y is recursively enumerable. Hence, by the simple iteration theorem there is a recursive function $t(y)$ such that for all x and y, $x \in \omega_{t(y)}$ iff $R_y(x, y)$.

Now, take any recursively enumerable relation $R(x, y)$. Then the relation $R(x, t(y))$ is also a recursively enumerable relation between x and y, hence has some index m. Thus $R_m(x, y)$ iff $R(x, t(y))$. Now comes the great trick: . . . .

Problem 6. Finish the proof.

As a consequence of Theorem 5, we have:

Theorem 6. *For any recursive function* $f(x)$ *there is a number* n *such that* $\omega_{f(n)} = \omega_n$.

Problem 7. (a) Prove Theorem 6. (b) Now prove the other four "believe it or nots".

Discussion. We have derived Theorem 6 as a corollary of Theorem 5. One can alternatively first prove Theorem 6 directly, and then obtain Theorem 5 as a corollary of Theorem 6 as follows: Assume Theorem 6. Now consider a recursively enumerable relation $R(x, y)$. By the iteration theorem, there is a recursive function $f(x)$ such that for all x and y,

$$x \in \omega_{f(y)} \text{ iff } R(x,y).$$

But by Theorem 6, there is a number n such that $\omega_{f(n)} = \omega_n$. Thus for all x, $x \in \omega_n$ iff $x \in \omega_{f(n)}$, which is true iff $R(x,n)$. Thus $x \in \omega_n$ iff $R(x,n)$.

Unsolvable Problems and Rice's Theorem

Some of the aforementioned believe it or nots may seem to be somewhat frivolous applications of the weak recursion theorem. We now turn to a far more significant application.

Let us call a property $P(n)$ *solvable* if the set of all n having the property is a recursive set. For example, suppose $P(n)$ is the property that ω_n is a finite set. To ask whether this property is solvable is to ask whether the set of all indices of all finite sets is recursive.

Here are some typical properties $P(n)$ that have come up in the literature:

(1) ω_n is an infinite set.
(2) ω_n is empty.
(3) ω_n is recursive.
(4) $\omega_n = \omega_1$.
(5) ω_n contains 1.
(6) The relation $R_n(x,y)$ is single-valued, i.e. for each x there is at most one y such that $R_n(x,y)$.
(7) The relation $R_n(x,y)$ is functional, i.e. for each x there is exactly one y such that $R_n(x,y)$.
(8) All numbers are in the domain of $R_n(x,y)$, i.e. for each (positive integer) x there is at least one y such that $R_n(x,y)$.
(9) All numbers are in the range of $R_n(x,y)$, i.e. for each (positive integer) y there is a least one x such that $R_n(x,y)$.

For each of these nine properties, we can ask if it is solvable. These are typical questions of recursion theory that have been answered by individual methods. Rice's theorem, to which we now turn, answers a host of such questions in one fell swoop.

A number set A is called *extensional* if it contains with any index of a recursively enumerable set all other indices of it as well — in other words, for every i and j, if $i \in A$ and $\omega_i = \omega_j$, then $j \in A$.

Theorem R. [Rice's Theorem] *The only recursive extensional sets are* $\mathbb{N}$ *and* $\emptyset$.

We will soon see how Rice's theorem [Rice, 1953] is relevant to the nine properties mentioned above, but first let us go to its proof. We first need the following:

Lemma. *For any recursive set A other than $\mathbb{N}$ and $\emptyset$, there is a recursive function $f(x)$ such that for all n, $n \in A$ iff $f(n) \notin A$.*

Problem 8. (a) Prove the above lemma. [**Hint:** If A is neither $\mathbb{N}$ nor $\emptyset$, then there is at least one number $a \in A$ and at least one number $b \notin A$. Use these two numbers a and b to define the function $f(x)$. (b) Now prove Rice's Theorem. [**Hint:** Use the function $f(x)$ of the Lemma and then apply Theorem 6 to the function $f(x)$.]

Applications

Let us consider the first of the aforementioned nine properties $P(n)$, namely that ω_n is infinite. Obviously, if ω_i is infinite and $\omega_i = \omega_j$, then ω_j is also infinite. Thus the set of the indices of all infinite sets is extensional. Also, there is at least one n such that ω_n is infinite and at least one n such that ω_n is not infinite. Hence by Rice's Theorem, the set of the indices of all infinite sets is not recursive.

The same argument applies to all the other eight cases, so that none of them are solvable!

The Strong Recursion Theorem

By the weak recursion theorem, for any number i, there is a number n such that $\omega_n = \{x : R_i(x, n)\}$. Given the number i, can we effectively find such an n as a recursive function of i? That is, is there a recursive function $\varphi(i)$ such that for every number i, $\omega_{\varphi(i)} = \{x : R_i(x, \varphi(i))\}$? The answer is *yes*, and this is Myhill's famous fixed point theorem, which is a special case of the following:

Theorem 7. [Strong Recursion Theorem] *For any recursively enumerable relation $M(x, y, z)$, there is a recursive function $\varphi(i)$ such that for all i,*
$$\omega_{\varphi(i)} = \{x : M(x, i, \varphi(i))\}.$$

We will give two proofs of this important theorem. The first is along traditional lines, and requires two applications of the iteration theorem. The second proof is along different lines, and requires only one application of the iteration theorem.

Here is the beginning of the first proof (to be finished by the reader in Problem 9(a)).

As in the proof of the weak recursion theorem, by the (simple) iteration theorem, for every y, there is a recursive function $d(y)$ such that $\omega_{d(y)} = \{x : R_y(x, y)\}$. Now, given a recursively enumerable relation $M(x, y, z)$, consider the recursively enumerable relation $M(x, z, d(y))$ as a relation among x, y and z. By another application of the iteration theorem, there is a recursive function $t(i)$ such that for all i, x and y, $R_{t(i)}(x, y)$ iff $M(x, i, d(y))$. Then ....

For the second proof (to be finished by the reader – Problem 9 (b) below), let $S(x, y, z)$ be the recursively enumerable relation $R_z(x, y, z)$. By the iteration theorem, there is a recursive function $t(y, z)$ such that for all y and z, $\omega_{t(y,z)} = \{x : S(x, y, z)\}$, and thus $\omega_{t(y,z)} = \{x : R_z(x, y, z)\}$.

Now, given a recursively enumerable relation $M(x, y, z)$, the relation $M(x, y, t(y, z))$ is recursively enumerable, hence has an index h. Thus $R_h(x, y, z)$ iff $M(x, y, t(y, z))$. Then ....

Problem 9. (a) Finish the first proof. (b) Finish the second proof.

As a special case of Theorem 7, we have:

Theorem 8. [Myhill's Fixed Point Theorem] *There is a recursive function* $\varphi(y)$ *such that for all* y, $\omega_{\varphi(y)} = \{x : R_y(x, \varphi(y))\}$.

Problem 10. Why is this a special case?

Creative, Productive, Generated and Universal Sets

To say that a set A is not recursively enumerable is to say that given any recursively enumerable subset ω_i of A there is a number n in A but not in ω_i — a witness, so to speak that ω_i is not the whole of A. A set A is called *productive* if there is a recursive function $\varphi(x)$ — called a *productive function for* A — such that for every number i, if ω_i is a subset of A, then $\varphi(i) \in A - \omega_i$ [i.e. $\varphi(i)$ is in A, but not in ω_i].

A set A is called *co-productive* if its complement $\overline{A}$ is productive. This is equivalent to the condition that there is a recursive function $\varphi(x)$ — called a *co-productive function for* A — such that for every number i, if ω_i is disjoint from A, then $\varphi(i)$ lies outside A and outside ω_i.

A set A is called *creative* if it is co-productive and recursively enumerable. Thus a creative set A is one which is recursively enumerable, and is such that there is a recursive function $\varphi(i)$ — called a *creative function for* A — such that for any number i, if ω_i is disjoint from A, then $\varphi(i)$ lies outside both A and ω_i (it is a witness to the fact that $\overline{A}$ is not recursively enumerable).

Post's Sets C and K

A simple example of a creative set is Post's set C, the set of all numbers x such that $x \in \omega_x$. Thus for any number i, $i \in C$ iff $i \in \omega_i$. If ω_i is disjoint from C, then i is outside both ω_i and C, and therefore the identity function $I(x)$ is a creative function for C.

A less immediate example of a creative set is the set K of all numbers $\partial(x, y)$ such that $x \in \omega_y$. The set K was also introduced by Post [1943] and is sometimes called the *complete set*. The proof that K is creative is less immediate than the proof for C, and involves use of the iteration theorem, as we will see.

Complete Productivity and Creativity

It is also correct to say that a set A is not recursively enumerable iff for every recursively enumerable set ω_i, whether a subset of A or not, there is a number j such that $j \in A$ iff $j \notin \omega_i$ — thus attesting to the fact that $A \neq \omega_i$. By a *completely productive function for* A is meant a recursive function $\varphi(x)$ such that for every number i, $\varphi(i) \in A$ iff $\varphi(i) \notin \omega_i$. A set A is called *completely productive* if there is a completely productive function for A.

A set A is called *completely creative* if A is recursively enumerable and there is a completely productive function $\varphi(x)$ for the complement $\overline{A}$ of A;

and such a function is called a *completely creative function for A*. Thus a completely creative function for A is a recursive function $\varphi(x)$ such that for every number i, $\varphi(i) \in A$ iff $\varphi(i) \in \omega_i$.

Post's set C of all x such that $x \in \omega_x$ is obviously not only creative, but completely creative, since the identity function is a completely creative function for C. Therefore a completely creative set exists. Post apparently did not capitalize on the fact that C is not only creative, but completely creative. I hazard a guess that if he had capitalized on this fact, then many results in recursion theory would have come to light sooner.

We will see later that in fact any productive set is completely productive, and hence that any creative set is completely creative. This important result is due to John Myhill [1955], and will be seen to be a consequence of the strong recursion theorem.

Generative Sets

Many important results about completely creative sets do not use the fact that the set is recursively enumerable, and so we define a set A to be *generative* if there is a recursive function $\varphi(x)$ — called a *generative function for A* — such that for all numbers i, $\varphi(i) \in A$ iff $\varphi(i) \in \omega_i$. Thus a generative set is like a completely creative set except that it is not necessarily recursively enumerable. A set A is generative iff its complement $\overline{A}$ is completely productive. Our emphasis will be on generative sets.

Many-One Reducibility

By a *(many-one) reduction of a set A to a set B* is meant a recursive function $f(x)$ such that $A = f^{-1}(B)$; i.e. for all numbers x, $x \in A$ iff $f(x) \in B$. A set A is said to be (many-one) reducible to B if there is a (many-one) reduction of A to B. There are other important types of reducibility that occur in the literature of recursion theory, but throughout this chapter, *reducible* will mean *many-one reducible*.

Theorem 9. *If A is reducible to B and A is generative, then B is generative.*

It will be useful to state and prove Theorem 9 in a more specific form: For any recursively enumerable relation $R(x, y)$, by the iteration theorem there is a recursive function $t(y)$ such that $\omega_{t(i)} = \{x : R(x,i)\}$. Such a function $t(y)$ will be called an *iterative function for the relation* $R(x,y)$. We will now state and prove Theorem 9 in the more specific form:

Theorem 9*. *If $f(x)$ reduces A to B and $\varphi(x)$ is a generative function for A, and $t(y)$ is an iterative function for the relation $f(x) \in \omega_y$, then $f(\varphi(t(x)))$ is a generative function for B.*

Problem 11. Prove Theorem 9*.

Universal Sets

A set A is called *universal* if every recursively enumerable set is reducible to A.

Post's complete set K is obviously universal.

Problem 12. Why?

Theorem 10. *Every universal set is generative.*

Problem 13. Prove Theorem 10. [**Hint:** Use Theorem 9.]

Uniformly Universal Sets

For any function $f(x,y)$ of two arguments, and for any number i, by f_i is meant the function which assigns to each x the number $f(x,i)$. Thus $f_i(x) = f(x,i)$.

A set A is called *uniformly universal under a recursive function* $f(x,y)$ if for all i and x, $x \in \omega_i$ iff $f(x,i) \in A$; in other words, iff f_i reduces ω_i to A. And we say that A is uniformly universal if it is universal under some recursive function $f(x,y)$.

Theorem 11. *Every universal set is uniformly universal.*

Problem 14. Prove Theorem 11. [**Hint:** Use the fact that Post's set K is not only universal, but is also uniformly universal. Why?]

Exercise. Suppose A is uniformly universal under $f(x,y)$ and that $t(y)$ is an iterative function for the relation $f(x,x) \in \omega_y$. Show that the function $f(t(x),t(x))$ is a generative function for A.

Remarks. The above, together with the fact that every universal set is uniformly universal, yields another proof that every universal set is generative, and unlike our previous proof of this fact, make no appeal to the existence of a previously constructed recursively enumerable generative set (for instance, Post's set C). Actually, the two proofs can be seen to yield two different generative functions for Post's set K.

Having shown that every universal set is generative, we now wish to show that, conversely, every generative set is universal, and thus that a set is generative if and only if it is universal.

To begin with, it is pretty obvious that if B is a generative set under $\varphi(x)$ (i.e. if $\varphi(x)$ is a generative function for B), then, for every i:
(1) If $\omega_i = \mathbb{N}$, then $\varphi(i) \in B$ [$\mathbb{N}$ is the set of positive integers];
(2) If $\omega_i = \emptyset$, then $\varphi(i) \notin B$ [$\emptyset$ is the empty set].

Problem 15. Why is this so?

Problem 16. Prove that for every recursively enumerable set A, there is a recursive function $t(y)$ such that for every number i:
(1) If $i \in A$, then $\omega_{t(i)} = \mathbb{N}$;
(2) If $i \notin A$, then $\omega_{t(i)} = \emptyset$. [Hint: Define $M(x, y)$ iff $y \in A$. The relation $M(x, y)$ is recursively enumerable, hence by the iteration theorem there is a recursive function $t(y)$ such that for all i, $\omega_{t(i)} = \{x : M(x, i)\}$.]

Theorem 12. *Every generative set is universal.*

Problem 17. Prove Theorem 12. [**Hint:** This follows fairly easily from Problems 15 and 16.]

We next wish to show that every co-productive set is generative. Then, by Theorem 12, we get Myhill's result that every co-productive set (and hence every creative set) is universal. [We recall that a set is creative iff it is both co-productive and recursively enumerable.]

Weak Co-productivity

We recall that a set A is said to be co-productive under a recursive function $\varphi(x)$ if for every number i such that ω_i is disjoint from A, the number $\varphi(i)$ lies outside both A and ω_i. This obviously implies the following weaker condition:

C_1: For every i such that ω_i is disjoint from A *and such that* ω_i *contains at most one member*, the number $\varphi(i)$ lies outside both A and ω_i.

Amazingly enough, this weaker condition on a recursive function is enough to ensure that A is generative (and hence universal).

This weaker condition C_1 implies the following even weaker condition:

C_2: For every number i,

    (1) If $\omega_i = \emptyset$. Then $\varphi(i) \notin A$;
    (2) If $\omega_i = \{\varphi(i)\}$, then $\varphi(i) \in A$.

Problem 18. Why does C_1 imply C_2?

We will say that A is *weakly co-productive under* $\varphi(x)$ if $\varphi(x)$ is recursive and condition C_2 holds.

Theorem 13. [After Myhill] *If A is co-productive — or even weakly co-productive — then A is generative.*

To prove Theorem 13, we need the following lemma:

Lemma. *For any recursive function $\varphi(x)$ there is a recursive function $t(y)$ such that for all y,*
$$\omega_{t(y)} = \omega_y \cap \{\varphi(t(y))\}.$$

Problem 19. Prove the above lemma. [**Hint:** Let $M(x, y, z)$ be the recursively enumerable relation $x \in \omega_y \wedge x = \varphi(z)$. Then use the strong recursion theorem, Theorem 7.]

Problem 20. Now prove Theorem 13. [**Hint:** Suppose $\varphi(x)$ is a weakly co-productive function for A. Let $t(y)$ be a recursive function related to $\varphi(x)$ as in the above lemma. Then show that the function $\varphi(t(y))$ is a generative function for A.]

From Theorems 13 and 12 we have:

Theorem 14. [Myhill] *If A is productive, or even weakly productive, then A is universal.*

Discussion

One can go directly from co-productivity (or even weak co-productivity) to universality as did Myhill. Briefly, the argument is the following: Suppose

B is weakly co-productive under $\varphi(x)$ and that A is a set that we wish to reduce to B. Applying the strong recursion theorem to the recursively enumerable relation $y \in A \wedge x = \varphi(z)$ [as a relation among x, y and z], we have a recursive function $t(y)$ such that for all y, if $y \in A$, then $\omega_{t(y)} = \{\varphi(y)\}$, and if $y \notin A$, then $\omega_{t(y)} = \emptyset$. Then the function $\varphi(t(x))$ can be seen to reduce A to B. The reader should find it a profitable exercise to fill in the details.

Recursive Isomorphism

A set A is said to be *recursively isomorphic* to a set B if there is a recursive 1-1 function from A *onto* B. Such a function $\varphi(x)$ is called a *recursive isomorphism* from A to B, and A is said to be recursively isomorphic to B under $\varphi(x)$, if $\varphi(x)$ is a recursive isomorphism from A to B.

If A is recursively isomorphic to B under $\varphi(x)$, then obviously B is recursively isomorphic to A under the inverse function $\varphi^{-1}(x)$. [We recall that $\varphi^{-1}(x) = y$ iff $\varphi(y) = x$.]

Exercise. Suppose A is recursively isomorphic to B. Prove that if A is recursively enumerable, (recursive, productive, generative, creative, universal) then B is respectively recursively enumerable (recursive, productive, generative, creative, universal).

Myhill [1955] proved the celebrated result that any two creative sets are recursively isomorphic.

Bonus Question. Is it possible for a universal set to be recursive?

Solutions to the Problems of Chapter 5

1. (a) We showed in the last chapter that the relation $g(x) = y$ (i.e. $x_0 = y$) is recursively enumerable. Hence the relation $r(x, y) = z$ is recursively enumerable, since it is equivalent to

$$\exists w(g(y) = w \wedge xw = z).$$

Therefore, for any number n, the relation $r(n, x) = y$ [as a relation between x and y] is recursively enumerable. And so for any recursively enumerable set A, the condition $r(n, x) \in A$ [as a property of x] is recursively enumerable, since it is equivalent to

$\exists y(r(n,x) = y \wedge y \in A)$. In particular, since T_0 is recursively enumerable, the condition $\exists y(r(n,x) = y \wedge y \in T_0)$ is recursively enumerable, and this is the condition $x \in \omega_n$. Thus ω_n is recursively enumerable.

(b) Every recursively enumerable set A is represented in (U) by some predicate H. Thus for every number x, $x \in A$ iff $Hx \in T$. We let h be the Gödel number of H, and so for all x, $Hx \in T$ iff $r(h,x) \in T_0$, which is true iff $x \in \omega_h$. Thus $A = \omega_h$.

2. The relation $R(x,y)$ is represented in (U) by some predicate H, and so for all numbers i and x, $R(i,x)$ iff $Hicomx \in T$, the latter of which is true iff $hi_0cx_0 \in T_0$, where h, i_0, c, x_0 are the respective Gödel numbers of H, i, com, x (in dyadic notation). Thus $R(i,x)$ iff $r(hi_0c, x) \in T_0$, which is true iff $x \in \omega_{hi_0c}$. We thus take $t(y)$ to be hy_0c, and so $t(i) = hi_0c$, so that now $R(i,x)$ iff $x \in \omega_{t(i)}$.

3. Given a recursively enumerable relation $R(x_1,\ldots,x_m,y_1,\ldots,y_n)$, we use the recursively enumerable relation $M(x,y)$, which yields, for all numbers $x_1,\ldots,x_m,i_1,\ldots,i_n$,
   (1) $R(x_1,\ldots,x_m,i_1,\ldots,i_n)$ iff $M(\delta_m(x_1,\ldots,x_m),\delta_n(i_1,\ldots,i_n))$.
   We now apply Theorem 1 to the relation $M(x,y)$, and so there is a recursive function $t(i)$ such that for every x and i we have $x \in \omega_{t(i)}$ iff $M(x,i)$. We now take $\delta_m(x_1,\ldots,x_m)$ for x and $\delta_n(i_1,\ldots,i_n)$ for i, and obtain
   (2) $\delta_m(x_1,\ldots,x_m) \in \omega_{t(\delta_n(i_1,\ldots,i_n))}$ iff $M(\delta_m(x_1,\ldots,x_m),\delta_n(i_1,\ldots,i_n))$.
   By (1) and (2) we have:
   (3) $R(x_1,\ldots,x_m,i_1,\ldots,i_n)$ iff $\delta_m(x_1,\ldots,x_m) \in \omega_{t(\delta_n(i_1,\ldots,i_n))}$ which is true iff $R_{t(\delta_n(i_1,\ldots,i_n))}(x_1,\ldots,x_m)$.
   We thus take $\varphi(i_1,\ldots,i_n)$ to be $t(\delta_n(i_1,\ldots,i_n))$ and obtain
   $$R(x_1,\ldots,x_m,i_1,\ldots,i_n) \text{ iff } R_{\varphi(i_1,\ldots,i_n)}(x_1,\ldots,x_m).$$

4. We will show that our recursively enumerable enumeration $\omega_1, \omega_2,\ldots,\omega_n,\ldots$ is maximal. Given a recursively enumerable enumeration $B_1, B_2,\ldots,B_n,\ldots$, let $R(x,y)$ be the recursively enumerable relation $x \in B_y$. By the simple iteration theorem, there is a recursive function $f(x)$ such that for all x and i, we have $x \in \omega_{f(i)}$ iff $R(x,i)$, which is true iff $x \in B_i$. Thus $x \in \omega_{f(i)}$ iff $x \in B_i$, and so $\omega_{f(i)} = B_i$.

5. Define $S(x_1,\ldots,x_m,y,y_1,\ldots,y_n)$ iff $R_y(x_1,\ldots,x_m,y_1,\ldots,y_n)$. The relation S is recursively enumerable (why?), and so by the iteration theorem, there is a recursive function $\varphi(i,i_1,\ldots,i_n)$

such that for all $x_1, \ldots, x_m, i, i_1, \ldots, i_n$: $R_{\varphi(i, i_1, \ldots, i_n)}(x_1, \ldots, x_m)$ iff $S(x_1, \ldots, x_m, i, i_1, \ldots, i_n)$ iff $R_i(x_1, \ldots, x_m, i_1, \ldots, i_n)$.

6. The great trick is to take m for y, and we thus obtain $R_m(x, m)$ iff $R(x, t(m))$. But we also have $R_m(x, m)$ iff $x \in \omega_{t(m)}$! Thus $x \in \omega_{t(m)}$ iff $R(x, t(m))$, and so $x \in \omega_n$ iff $R(x, n)$, where n is the number $t(m)$. [Clever, huh?]

7. (a) Given a recursive function $f(x)$ let $R(x, y)$ be the relation $x \in \omega_{f(y)}$. It is recursively enumerable [since it is equivalent to $\exists z(f(y) = z \wedge x \in \omega_z)$], so that by Theorem 5, there is some n such that for all x, $x \in \omega_n$ iff $R(x, n)$. But $R(x, n) \equiv x \in \omega_{f(n)}$, and so, for all x, $x \in \omega_n$ iff $x \in \omega_{f(n)}$. Thus $\omega_n = \omega_{f(n)}$.

   (b) Believe it or not:

   #1 is but a special case of (a), namely when $f(x) = x + 1$.

   #2 is just Theorem 6!

   #3 is but the special case of #4 in which $f(x) = x$. And so we must show #4: For $f(x)$ a recursive function, the relation $x = f(y)$ is recursively enumerable. Hence by Theorem 5, there is some n such that for all $x, x \in \omega_n$ iff $x = f(n)$. This means that $f(n)$ is the one and only member of ω_n, and so $\omega_n = \{f(n)\}$.

   As for #5, let $R(x, y)$ be the relation $y \in \omega_x$. This relation is recursively enumerable, and so there is a number n such that for all x, $x \in \omega_n$ iff $R(x, n)$, but $R(x, n)$ iff $n \in \omega_x$, and so $x \in \omega_n$ iff $x \in \omega_x$.

8. (a) We are given that A is recursive and that $A \neq \mathbb{N}$ and $A \neq \emptyset$. Since $A \neq \emptyset$, then there is at least one number a such that $a \in A$, and since $A \neq \mathbb{N}$ there is at least one number b such that $b \notin A$. Define $f(x)$ to be b if $x \in A$ and to be a if $x \notin A$. Since A is recursive, then A and its complement $\overline{A}$ are both recursively enumerable. Hence the function $f(x)$ is recursive, since the relation $f(x) = y$ is equivalent to the recursively enumerable relation $(x \in A \wedge y = b) \vee (x \in \overline{A} \wedge y = a)$. If $x \in A$, then $f(x) \notin A$ [since $f(x) = b$], and if $x \notin A$, then $f(x) \in A$ [since $f(x) = a$]. Thus $x \in A$ iff $f(x) \notin A$.

   (b) Let A be a recursive set that is neither $\mathbb{N}$ nor $\emptyset$. Take $f(x)$ as in the lemma. By Theorem 6, there is some n such that $\omega_n = \omega_{f(n)}$. Thus n and $f(n)$ are indices of the same recursively enumerable set, yet one of them is in A and the other isn't. Therefore A is not extensional.

9. (a) If we take $t(i)$ for y in the statement $R_{t(i)}(x,y)$ iff $M(x,i,d(y))$ we obtain $R_{t(i)}(x,t(i))$ iff $M(x,i,d(t(i)))$. But it is also true that $\omega_{d(t(i))} = \{x : R_{t(i)}(x,t(i))\}$ [this is true because, for any n, $\omega_{d(n)} = \{x : R_n(x,n)\}$]. Thus $\omega_{d(t(i))} = \{x : M(x,i,d(t(i)))$. We now take $\varphi(i)$ to be $d(t(i))$, yielding $\omega_{\varphi(i)} = \{x : M(x,i,\varphi(i))\}$.

   (b) As noted, $R_h(x,y,z)$ iff $M(x,y,t(y,z))$. We take h for z and so $R_h(x,y,h)$ iff $M(x,y,t(y,h))$. But since $\omega_{t(y,h)} = \{x : R_h(x,y,h)\}$, it follows that $\omega_{t(y,h)} = \{x : M(x,y,(t(y,h)))\}$. We thus take $\varphi(y)$ to be $t(y,h)$, yielding $\omega_{\varphi(y)} = \{x : M(x,y,\varphi(y))\}$.

10. This is the case that $M(x,y,z)$ is the relation $R_y(x,z)$.

11. Since f reduces A to B, then for all x, $f(x) \in B$ iff $x \in A$. For any number i we take $\varphi(t(i))$ for x and we obtain:
    (1) $f(\varphi(t(i))) \in B$ iff $\varphi(t(i)) \in A$.
    Also, $\varphi(t(i)) \in A$ iff $\varphi(t(i)) \in \omega_{t(i)}$ [since $\varphi(x)$ is a generative function for A], so we have that:
    (2) $f(\varphi(t(i))) \in B$ iff $\varphi(t(i)) \in \omega_{t(i)}$.
    Next, since $t(y)$ is an iterative function for the relation $f(x) \in \omega_y$, then for all x and i we have $x \in \omega_{t(i)}$ iff $f(x) \in \omega_i$. We now take $\varphi(t(i))$ for x and obtain:
    (3) $\varphi(t(i)) \in \omega_{t(i)}$ iff $f(\varphi(t(i))) \in \omega_i$.
    By (2) and (3), we see that $f(\varphi(t(i))) \in B$ iff $f(\varphi(t(i))) \in \omega_i$, and thus $f(\varphi(t(x)))$ is a generative function for B.

12. For any number i, $x \in \omega_i$ iff for all $x, \delta(x,i) \in K$. Let φ_i be the function that assigns to each x the number $\delta(x,i)$. Thus $\varphi_i(x) = \delta(x,i)$. Consequently, $x \in \omega_i$ iff $\varphi_i(x) \in K$, and so φ_i reduces ω_i to K.

13. Suppose A is universal. Post's set C is recursively enumerable and generative. Since C is recursively enumerable and A is universal, then C is reducible to A. Then, since C is generative, so is A [by Theorem 9].

14. Post's set K is obviously uniformly universal under the recursive function $\delta(x,y)$. Now, suppose A is universal. Then K is reducible to A under some recursive function $f(x)$. Then A must be uniformly universal under the function $f(\delta(x,y))$, because for any numbers i and x, $x \in \omega_i$ iff $\delta(x,i) \in K$, which is true iff $f(\delta(x,i)) \in A$. Thus $x \in \omega_i$ iff $f(\delta(x,i)) \in A$.

15. We are given that $\varphi(x)$ is a generative function for B. Thus for every number i,

$$\text{(a)} \quad \varphi(i) \in \omega_i \text{ iff } \varphi(i) \in B.$$

(1) Suppose $\omega_i = \mathbb{N}$. Then by (a), $\varphi(i) \in \mathbb{N}$ iff $\varphi(i) \in B$, but $\varphi(i)$ *is* in $\mathbb{N}$, so $\varphi(i) \in B$.

(2) Suppose $\omega_i = \emptyset$. Then by (a), $\varphi(i) \in \emptyset$ iff $\varphi(i) \in B$, but $\varphi(i) \notin \emptyset$, so $\varphi(i) \notin B$.

16. For all x, we have $M(x, y)$ iff $y \in A$. We are letting $t(y)$ be an iterative function for the relation $M(x, y)$, i.e. t is a recursive function such that for every x and i: $x \in \omega_{t(i)}$ iff $M(x, i)$, which is true iff $i \in A$. Thus $x \in \omega_{t(i)}$ iff $i \in A$.

(1) If $i \in A$, then for every x, $x \in \omega_{t(i)}$, which means that $\omega_{t(i)} = \mathbb{N}$.

(2) If $i \notin A$, then for every x, $x \notin \omega_{t(i)}$, which means that $\omega_{t(i)}$ is the empty set $\emptyset$.

17. Suppose B is a generative set. We wish to show that B is universal. Let $\varphi(x)$ be a generative function for B. Consider any recursively enumerable set A which we wish to reduce to B. Let $t(y)$ be a recursive function related to A as in Problem 16. We assert that the function $\varphi(t(x))$ reduces A to B.

(1) Suppose $i \in A$. Then $\omega_{t(i)} = \mathbb{N}$ [by Problem 16(1)]. Hence $\omega_{\varphi(t(i))} \in B$ [by Problem 15(a), taking $t(i)$ for i].

(2) Suppose $i \notin A$. Then $\omega_{t(i)} = \emptyset$ [by Problem 16(2)]. Hence $\omega_{\varphi(t(i))} \notin B$ [by Problem 15(b), taking $t(i)$ for i].

Thus if $i \in A$, then $\varphi(t(i)) \in B$, and if $i \notin A$, then $\varphi(t(i)) \notin B$, or, what is the same thing, if $\varphi(t(i)) \in B$ then $i \in A$. Thus $i \in A$ iff $\varphi(t(i)) \in B$, and so the function $\varphi(t(x))$ reduces A to B.

18. Assume condition C_1.

(1) Suppose $\omega_i = \emptyset$. Obviously $\emptyset$ is disjoint from A, and so by C_1, the number $\varphi(i) \notin A$ [also, $\varphi(i) \notin \omega_i$].

(2) Suppose $\omega_i = \{\varphi(i)\}$. If ω_i were disjoint from A, then by C_1, we would have $\varphi(i) \notin \omega_i$, and thus $\varphi(i) \notin \varphi(i)\}$, which is absurd. Thus $\varphi(i) \in A$.

19. Define $M(x, y, z)$ to be $x \in \omega_y \wedge x = \varphi(z)$. By the strong recursion theorem (Theorem 7), there is a recursive function $t(y)$ such that
$$\omega_{t(y)} = \{x : M(x, y, (t(y)))\}$$
$$= \{x : x \in \omega_y \wedge x = \varphi(t(y))\},$$
$$= \{x : x \in \omega_y\} \cap \{x : x = \varphi(t(y))\},$$
$$= \omega_y \cap \{\varphi(t(y))\}.$$

20. We want to show that for any number y, $\varphi(t(y)) \in \omega_y$ iff $\varphi(t(y)) \in A$.

   (1) Suppose $\varphi(t(y)) \in \omega_y$. Then $\omega_y \cap \{\varphi(t(y))\} = \{\varphi(t(y))\}$. But also $\omega_y \cap \{\varphi(t(y))\} = \omega_{t(y)}$ [by the Lemma]. Then $\varphi(t(y)) \in A$ [by condition C_2, taking $t(y)$ for i, since A is weakly co-productive under $\varphi(x)$]. Thus if $\varphi(t(y)) \in \omega_y$, then $\varphi(t(y)) \in A$.

   (2) Suppose $\varphi(t(y)) \notin \omega_y$. Then $\omega_y \cap \{\varphi(t(y))\} = \emptyset$. Again by the Lemma, $\omega_y \cap \{\varphi(t(y))\} = \omega_{t(y)}$. Thus $\omega_{t(y)} = \emptyset$. Again by condition C_2, taking $t(y)$ for i, $\varphi(t(y)) \notin A$. Thus if $\varphi(t(y)) \notin \omega_y$, then $\varphi(t(y)) \notin A$, or, equivalently if $\varphi(t(y)) \in A$, then $\varphi(t(y)) \in \omega_y$, and thus $\varphi(t(y))$ is a generative function for A.

Answer to the Bonus Question. Of course not! A universal set is generative and a generative set A cannot be recursive, for if it were, its complement $\overline{A}$ would be recursively enumerable, hence there would be a number n such that $n \in A$ iff $n \in \overline{A}$, which is impossible.

Chapter 6

Doubling Up

For my intended applications to the mathematical systems of the next chapter, we will need double analogues of earlier results.

Double Recursion Theorems

In preparation for the double recursion theorems we will consider here, we note that by the iteration theorem, there is a recursive function d such that for all y and z, $\omega_{d(y,z)} = \{x : R_y(x, y, z)\}$. This function will play a key role, and I shall call it the *double diagonal function*.

Theorem 1. [Weak Double Recursion Theorem] *For any two recursively enumerable relations $M_1(x, y, z)$ and $M_2(x, y, z)$, there are numbers a and b such that:*
(1) $\omega_a = \{x : M_1(x, a, b)\}$;
(2) $\omega_b = \{x : M_2(x, a, b)\}$.

Problem 1. Prove Theorem 1. [**Hint:** Use the double diagonal function d and consider the relations $M_1(x, d(y, z), d(z, y))$ and $M_2(x, d(z, y), d(y, z))$.]

The Strong Double Recursion Theorem

We now wish to show that in Theorem 1, the numbers a and b can be found as recursive function of the indices i and j of the relations $M_1(x, y, z)$ and $M_2(x, y, z)$. This is Theorem 3 below, which will soon be seen to be a consequence of:

Theorem 2. [Strong Double Recursion Theorem] *For any recursively enumerable relations* $M_1(x, y_1, y_2, z_1, z_2)$ *and* $M_2(x, y_1, y_2, z_1, z_2)$, *there are recursive functions* $\varphi_1(i,j)$ *and* $\varphi_2(i,j)$ *such that for all numbers* i *and* j:

(1) $\omega_{\varphi_1(i,j)} = \{x : M_1(x, i, j, \varphi_1(i,j), \varphi_2(i,j))\}$;

(2) $\omega_{\varphi_2(i,j)} = \{x : M_2(x, i, j, \varphi_1(i,j), \varphi_2(i,j))\}$.

Problem 2. Prove Theorem 2. [**Hint:** We again use the double diagonal function d. By the iteration theorem there are recursive functions $t_1(i,j)$ and $t_2(i,j)$ such that for all numbers i, j, x, y, z:

(1) $R_{t_1(i,j)}(x, y, z)$ iff $M_1(x, i, j, d(y,z), d(z,y))$;

(2) $R_{t_2(i,j)}(x, y, z)$ iff $M_2(x, i, j, d(z,y), d(y,z))$.]

As an obvious corollary of Theorem 2 we have:

Theorem 3. *For any two recursively enumerable relations* $M_1(x, y, z_1, z_2)$ *and* $M_2(x, y, z_1, z_2)$, *there are recursive functions* $\varphi_1(i,j)$ *and* $\varphi_2(i,j)$ *such that for all numbers* i *and* j:

(1) $\omega_{\varphi_1(i,j)} = \{x : M_1(x, i, \varphi_1(i,j), \varphi_2(i,j))\}$;

(2) $\omega_{\varphi_2(i,j)} = \{x : M_2(x, j, \varphi_1(i,j), \varphi_2(i,j))\}$.

A special case of Theorem 3 is:

Theorem 4. [Strong Double Myhill Theorem] *There are recursive functions* $\varphi_1(i,j)$ *and* $\varphi_2(i,j)$ *such that for all numbers* i *and* j:

(1) $\omega_{\varphi_1(i,j)} = \{x : R_i(x, \varphi_1(i,j), \varphi_2(i,j))\}$;

(2) $\omega_{\varphi_2(i,j)} = \{x : R_j(x, \varphi_1(i,j), \varphi_2(i,j))\}$.

Problem 3. (a) Why is Theorem 3 a corollary of Theorem 2? (b) Why is Theorem 4 a special case of Theorem 3?

Discussion

The proof we have given of the strong double recursion theorem involved three applications of the iteration theorem. Actually, the theorem can be proved using only one application of the iteration theorem, and that can be done in more than one way.

Several workers in the field of recursion theory have found a way of deriving the *weak* double recursion theorem as a consequence of the strong (single) recursion theorem, and I have subsequently found a way of deriving the *strong* double recursion theorem as a consequence of the strong (single) recursion theorem.

All this can be found in a more in-depth study of recursion theorems in my book *Recursion Theory for Metamathematics* [Smullyan, 1993], in which several other variants of the recursion theorem are also treated.

More Doubling Up

We shall say that an ordered pair (A_1, A_2) is *reducible* to the ordered pair (B_1, B_2) under a recursive function $f(x)$ if $f(x)$ simultaneously reduces A_1 to B_1 and A_2 to B_2. This means that $A_1 = f^{-1}(B_1)$ and $A_2 = f^{-1}(B_2)$; in other words, it means that for every number x:

(1) If $x \in A_1$, then $f(x) \in B_1$;

(2) If $x \in A_2$, then $f(x) \in B_2$;

(3) If $x \notin A_1 \cup A_2$, then $f(x) \notin B_1 \cup B_2$.

If (1) and (2) above both hold, without (3) necessarily holding, we say that (A_1, A_2) is *semi-reducible* to the ordered pair (B_1, B_2) under the recursive function $f(x)$.

We shall call a pair (U_1, U_2) *doubly universal* if every disjoint pair (A, B) of recursively enumerable sets is reducible to (U_1, U_2). We shall call a pair (U_1, U_2) *semi-doubly universal* if every disjoint pair (A, B) of recursively enumerable sets is semi-reducible to (U_1, U_2).

We shall call a disjoint pair (A, B) *doubly generative* under a recursive function $\varphi(x, y)$ if, for every disjoint pair (ω_i, ω_j) of recursively enumerable sets, the following holds:

(1) $\varphi(i, j) \in A$ iff $\varphi(i, j) \in \omega_i$;

(2) $\varphi(i, j) \in B$ iff $\varphi(i, j) \in \omega_j$.

In this chapter we will prove that a pair (A, B) is doubly generative iff it is doubly universal.

We shall call a disjoint pair (A, B) *doubly co-productive* under a recursive function $g(x, y)$ if for every disjoint pair of recursively enumerable sets (ω_i, ω_j), if ω_i is disjoint from A and ω_j is disjoint from B, then the number $g(i, j)$ is outside all four sets ω_i, ω_j, A and B.

It is obvious that if (A, B) is doubly generative, then (A, B) is doubly co-productive. The converse is also true, but this is not trivial. The proof requires something like the strong double recursion theorem.

All this is closely connected to the topic of effective inseparability, to which we now turn.

Effective Inseparability

First we must discuss a very useful principle: Consider two recursively enumerable sets A and B. They are both Σ_1 and so there are Σ_0 relations $R_1(x, y)$ and $R_2(x, y)$ such that for all x,

(1) $x \in A$ iff $\exists y R_1(x, y)$;

(2) $x \in B$ iff $\exists y R_2(x, y)$.

Now define the sets A' and B' by the following conditions:

$$x \in A' \text{ iff } \exists y (R_1(x, y) \wedge (\forall z \leq y)(\sim R_2(x, y))).$$

$$x \in B' \text{ iff } \exists y (R_2(x, y) \wedge (\forall z \leq y)(\sim R_1(x, y))).$$

The sets A' and B' are both recursively enumerable and disjoint.

Problem 4. Why are the sets A' and B' disjoint?

We thus have:

Separation Principle: For any two recursively enumerable sets A and B, there are disjoint recursively enumerable sets A' and B' such that $A - B \subseteq A'$ and $B - A \subseteq B'$.

A disjoint pair of sets (A, B) is said to be *recursively inseparable* if there is no *recursive* superset of A disjoint from B, or, what is the same thing, if there is no recursive superset of B disjoint from A (these are the same because if C is a recursive superset of B disjoint from A, its complement $\overline{C}$ is a recursive superset of A disjoint from B). A pair of sets (A, B) is said to be *recursively separable* if the pair is not recursively inseparable (i.e. there *does* exist a recursive superset of A disjoint from B or a recursive superset of B disjoint from A). Clearly a pair of sets (A, B) that is recursively separable must be disjoint.

Now, to say that (A, B) is recursively inseparable is equivalent to saying that for any disjoint recursively enumerable supersets ω_i and ω_j of A and B respectively, there is a number n outside both ω_i and ω_j [i.e. ω_i is not the complement of ω_j].

Problem 5. Why are the characterizations of recursive inseparability in the two preceding paragraphs equivalent?

A disjoint pair (A, B) is called *effectively inseparable* if there is a recursive function $\zeta(x, y)$ [the Greek letter ζ is read as "zeta"] — called an effective inseparability function for (A, B) — such that for any recursively enumerable disjoint supersets ω_i and ω_j of A and B respectively, the number $\zeta(i, j)$ lies outside both ω_i and ω_j [$\zeta(i, j)$ is a witness, so to speak, that ω_i is not the complement of ω_j].

We shall call a disjoint pair (A, B) *completely effectively inseparable* if there is a recursive function $\zeta(x, y)$ — called a completely effective inseparability function for (A, B) — such that for any recursively enumerable supersets ω_i and ω_j of A and B respectively, whether they are disjoint or not, the number $\zeta(i, j)$ is either in both ω_i and ω_j or in neither, in other words $\zeta(x, y) \in \omega_i$ iff $\zeta(x, y) \in \omega_j$. [If ω_i and ω_j happen to be disjoint, then $\zeta(x, y)$ is obviously outside both ω_i and ω_j, and so any completely effective inseparability function for (A, B) is also an effective inseparability function for (A, B).]

We will later show that the non-trivial result that if (A, B) is effectively inseparable, and A and B are both recursively enumerable, then (A, B) is completely effectively inseparable. The only proof of this that I know of uses the strong double recursion theorem.

Kleene's Construction

There are several ways to construct a completely effectively inseparable pair of recursively enumerable sets. The following way is due to Kleene [1952].

We shall call a recursive function $k(x, y)$ a *Kleene function* for a disjoint pair (A_1, A_2) if, for all numbers i and j,

(1) If $k(i, j) \in \omega_i - \omega_j$, then $k(i, j) \in A_2$;

(2) If $k(i, j) \in \omega_j - \omega_i$, then $k(i, j) \in A_1$.

Proposition 1. *If $k(x, y)$ is a Kleene function for a disjoint pair (A_1, A_2), then (A_1, A_2) is completely effectively inseparable under $k(x, y)$.*

Problem 6. Prove Proposition 1.

We shall call a disjoint pair (A_1, A_2) a *Kleene pair* if there is a Kleene function for (A_1, A_2).

By Proposition 1, any Kleene pair is completely effectively inseparable.

Theorem 5. *There exists a Kleene pair (K_1, K_2) of recursively enumerable sets.*

Problem 7. Prove Theorem 5. [**Hint**: Use the pairing function $\delta(x, y)$. Let A_1 be the set of all numbers $\delta(i, j)$ such that $\delta(i, j) \in \omega_j$ and let A_2 be the set of all numbers $\delta(i, j)$ such that $\delta(i, j) \in \omega_i$. Show that $\delta(x, y)$ is a Kleene function for the pair $(A_1 - A_2, A_2 - A_1)$. Then use the separation principle to get the recursively enumerable sets K_1 and K_2.]

From Theorem 5 and Proposition 1, we have:

Theorem 6. *There is a completely effectively inseparable pair of recursively enumerable sets, to wit, the Kleene pair* (K_1, K_2).

Proposition 2. *If* (A_1, A_2) *is a Kleene pair and if* (A_1, A_2) *is semireducible to* (B_1, B_2) *and* B_1 *is disjoint from* B_2, *then* (B_1, B_2) *is a Kleene pair.*

Problem 8. Prove Proposition 2. [**Hint**: Let $f(x)$ be a function that semi-reduces (A_1, A_2) to (B_1, B_2), and let $k(x, y)$ be a Kleene function for (A_1, A_2). By the iteration theorem, there is a recursive function $t(y)$ such that $\omega_{t(y)} = \{x : f(x) \in \omega_y\}$. Show that $\omega_{t(y)} = f^{-1}(\omega_y)$. From the functions $k(x, y)$, $f(x)$, $t(y)$ construct a Kleene function for (B_1, B_2).]

Theorem 7. *Every disjoint semi-doubly-universal pair is completely effectively inseparable.*

Problem 9. Why is Theorem 7 true?

Doubly Generative Pairs

We recall that a disjoint pair (A, B) is called *doubly generative* if there is a recursive function $\varphi(x, y)$ — called a doubly generative function for (A, B) — such that, for every disjoint pair (ω_i, ω_j), the following holds:

(1) $\varphi(i, j) \in A$ iff $\varphi(i, j) \in \omega_i$;
(2) $\varphi(i, j) \in B$ iff $\varphi(i, j) \in \omega_j$.

Problem 10. Prove that if (A, B) is called *doubly generative*, then A and B are each generative sets.

The next theorem is vital:

Theorem 8. *If* (A, B) *is completely effectively inseparable and* A *and* B *are both recursively enumerable, then* (A, B) *is doubly generative.*

Problem 11. Prove Theorem 8. [**Hint**: (A, B) is completely effectively inseparable under some recursive function $\zeta(x, y)$. By the iteration theorem, there are recursive functions $t_1(n), t_2(n)$ such that for all x and n:

(1) $\omega_{t_1(n)} = \{x : x \in \omega_n \vee x \in A\}$;

(2) $\omega_{t_2(n)} = \{x : x \in \omega_n \vee x \in B\}$;

Let $\varphi(i, j) = \zeta(t_1(j), t_2(i))$. Show that (A, B) is doubly generative under the function $\varphi(i, j)$.]

Since the Kleene pair (K_1, K_2) is completely effectively inseparable and K_1 and K_2 are both recursively enumerable, it follows from Theorem 8 that the pair (K_1, K_2) is doubly generative. Thus there exists a doubly generative pair of recursively enumerable sets.

Doubly Generative, Doubly Universal

Suppose (A_1, A_2) is a disjoint pair which is doubly generative under the function $g(x, y)$. Then, for all numbers i and j, the following three conditions hold:

C_1: If $\omega_i = \mathbb{N}$ and $\omega_j = \emptyset$, then $g(i, j) \in A_1$;

C_2: If $\omega_i = \emptyset$ and $\omega_j = \mathbb{N}$, then $g(i, j) \in A_2$;

C_3: If $\omega_i = \emptyset$ and $\omega_j = \emptyset$, then $g(i, j) \notin A_1 \cup A_2$.

Problem 12. Prove this.

Problem 13. Suppose (A_1, A_2) is a disjoint pair and that $g(x, y)$ is a recursive function such that conditions C_1 and C_2 above hold. Prove that for any disjoint pair (B_1, B_2) of recursively enumerable sets there are recursive functions $t_1(x), t_2(x)$ such that:

(1) (B_1, B_2) is semi-reducible to (A_1, A_2) under $g(t_1(x), t_2(x))$;

(2) If condition C_3 also holds, then (B_1, B_2) is reducible to (A_1, A_2) under $g(t_1(x), t_2(x))$. [**Hint**: Use the result of Problem 16 of the previous chapter.]

From Problems 12 and 13 immediately follows:

Theorem 9. *Every doubly generative pair is doubly universal.*

We now see that for a disjoint pair (A, B), being completely effectively inseparable, being doubly generative, and being doubly universal, are all one and the same thing. Now to complete the picture.

Double Co-productivity

We recall that a disjoint pair (A, B) is defined to be *doubly co-productive* under a recursive function $g(x, y)$ if for all i and j, if ω_i is disjoint from ω_j, and ω_i is disjoint from A and ω_j is disjoint from B, then $g(i, j)$ lies outside all the four sets ω_i, ω_j, A and B.

We claim that if (A, B) is *doubly co-productive* under a recursive function $g(x, y)$, then, for all numbers i and j, the following conditions hold:

D_0: If $\omega_i = \omega_j = \emptyset$, then $g(i, j) \notin A \cup B$;
D_1: If $\omega_i = \{g(i, j)\}$ and $\omega_j = \emptyset$, then $g(i, j) \in A$;
D_2: If $\omega_i = \emptyset$ and $\omega_j = \{g(i, j)\}$, then $g(i, j) \in B$.

Problem 14. Prove that if (A, B) is doubly co-productive under the recursive function $g(x, y)$, then conditions D_0, D_1 and D_2 hold.

We will say that (A, B) is called *weakly doubly co-productive* under the recursive function $g(x, y)$ if conditions D_0, D_1 and D_2 hold.

We now wish to prove:

Theorem 10. *If (A, B) is doubly co-productive, or even weakly doubly co-productive, under a recursive function $g(x, y)$, then (A, B) is doubly generative (and hence also doubly universal).*

For the proof of Theorem 10 we need:

Lemma. *For any recursive function $g(x, y)$, there are recursive functions $t_1(i, j), t_2(i, j)$ such that for all numbers all i and j,*
(1) $\omega_{t_1(i,j)} = \omega_i \cap \{g(t_1(i, j), t_2(i, j))\}$;
(2) $\omega_{t_2(i,j)} = \omega_j \cap \{g(t_1(i, j), t_2(i, j))\}$.

To prove this, we will need the strong double recursion theorem, or rather its corollary, Theorem 3.

Problem 15. Prove the above Lemma. [**Hint:** Let $M_1(x, y, z_1, z_2)$ be the recursively enumerable relation $x \in \omega_y \wedge x = g(z_1, z_2)$, and let $M_2(x, y, z_1, z_2)$ be the same relation. Then use Theorem 3.]

Problem 16. Now prove Theorem 10. [**Hint:** Suppose (A, B) is weakly doubly co-productive under $g(x, y)$. By the lemma, there are recursive functions $t_1(x, y), t_2(x, y)$ such that for all i and j,

(1) $\omega_{t_1(i,j)} = \omega_i \cap \{g(t_1(y_1,y_2), t_2(y_1,y_2))\}$;

(2) $\omega_{t_2(i,j)} = \omega_j \cap \{g(t_1(y_1,y_2), t_2(y_1,y_2))\}$.

Let $\varphi(x,y) = g(t_1(x,y)t_2(x,y))$, and show that (A,B) is doubly generative under $\varphi(x,y)$.]

Next we wish to prove:

Theorem 11. *If (A,B) is effectively inseparable and A and B are both recursively enumerable, then (A,B) is doubly generative (and hence doubly universal).*

Problem 17. Prove Theorem 11. [**Hint**: (a) First prove that for any recursively enumerable set A there is a recursive function $t(i)$ such that for all $i, \omega_{t(i)} = \omega_i \cup A$. (b) Now suppose that (A,B) is effectively inseparable under $g(x,y)$ and that A and B are both recursively enumerable sets. Let $t_1(i), t_2(i)$ be recursive functions such that for all i, $\omega_{t_1(i)} = \omega_i \cup A$ and $\omega_{t_2(i)} = \omega_i \cup B$. Let $g(x,y) = \zeta(t_1(y), t_2(x))$ and show that (A,B) is doubly generative under the function $g(x,y)$.]

Solutions to the Problems of Chapter 6

1. Let m be an index of the relation $M_1(x, d(y,z), d(z,y))$, and let n be an index of $M_2(x, d(z,y), d(y,z))$. Then:

   (1) $R_m(x,y,z)$ iff $M_1(x, d(y,z), d(z,y))$,

   (2) $R_n(x,y,z)$ iff $M_2(x, d(z,y), d(y,z))$.

   In (1), take m for y and n for z, and in (2), take n for y and m for z. As a result, we have:

   (1') $R_m(x,m,n)$ iff $M_1(x, d(m,n), d(n,m))$,

   (2') $R_n(x,n,m)$ iff $M_2(x, d(m,n), d(n,m))$.

   Now, $\omega_{d(m,n)} = \{x : R_m(x,m,n)\} = \{x : M_1(x, d(m,n), d(n,m))\}$ [by (1')]; and

   $\omega_{d(n,m)} = \{x : R_n(x,n,m)\} = \{x : M_2(x, d(m,n), d(n,m))\}$ [by (2')].

   We thus take $a = d(m,n)$ and $b = d(n,m)$, and have:

   $$\omega_a = \{x : M_1(x,a,b)\} \quad \text{and} \quad \omega_b = \{x : M_2(x,a,b)\}.$$

2. We define:

   $$\varphi_1(i,j) = d(t_1(i,j), t_2(i,j)) \quad \text{and}$$

   $$\varphi_2(i,j) = d(t_2(i,j), t_1(i,j)),$$

where $t_1(i,j)$ and $t_2(i,j)$ are the functions mentioned in the hint and $d(x,y)$ is the double diagonal function. We will see that these two functions work.

By the definitions of t_1 and t_2, we know that for all numbers i, j, x, y, z,

(1) $R_{t_1(i,j)}(x,y,z)$ iff $M_1(x,i,j,d(y,z),d(z,y))$;

(2) $R_{t_2(i,j)}(x,y,z)$ iff $M_2(x,i,j,d(z,y),d(y,z))$.

In (1), we take $t_1(i,j)$ for y and $t_2(i,j)$ for z, and in (2), we take $t_2(i,j)$ for y and $t_1(i,j)$ for z, obtaining:

(1') $R_{t_1(i,j)}(x,t_1(i,j),t_2(i,j))$ iff $M_1(x,i,j,\varphi_1(i_1,i_2),\varphi_2(i_1,i_2))$;

(2') $R_{t_2(i,j)}(x,t_2(i,j),t_1(i,j))$ iff $M_2(x,i,j,\varphi_1(i_1,i_2),\varphi_2(i_1,i_2))$.

Also:

(3) $\omega_{\varphi_1(i,j)} = \omega_{d(t_{1(i,j)},t_{2(i,j)})} = \{x : R_{t_1(i,j)}(x,t_{1(i,j)},t_{2(i,j)})\}$;

(4) $\omega_{\varphi_2(i,j)} = \omega_{d(t_{2(i,j)},t_{1(i,j)})} = \{x : R_{t_2(i,j)}(x,t_{2(i,j)},t_{1(i,j)})\}$.

From (3) and (1'), we have $\omega_{\varphi_1(i,j)} = \{x : M_1(i,j, \varphi_1(i,j), \varphi_2(i,j))\}$.
From (4) and (2'), we have $\omega_{\varphi_2(i,j)} = \{x : M_2(i,j, \varphi_1(i,j), \varphi_2(i,j))\}$.

3. (a) Given the recursively enumerable relations $M_1(x,y,z_1,z_2)$ and $M_2(x,y,z_1,z_2)$ define two new relations $M_1'(x,y_1,y_2,z_1,z_2)$ iff $M_1(x,y_1,z_1,z_2)$ and $M_2'(x,y_1,y_2,z_1,z_2)$ iff $M_2(x,y_2,z_1,z_2)$. Then apply Theorem 2 to the relations M_1' and M_2'.

   (b) This follows from Theorem 3 by taking $M_1(x,y,z_1,z_2)$ and $M_2(x,y,z_1,z_2)$ to both be the relation $R_y(x,z_1,z_2)$.

4. If $x \in A'$ and $x \in B'$, we get the following contradiction: There must be numbers n and m such that:

   (1) $R_1(x,n)$.

   (2) $R_2(x,m)$.

   (3) If $R_2(x,m)$, then $m > n$ [since $(\forall z \leq n) \sim R_2(x,z)$].

   (4) If $R_1(x,n)$, then $n > m$ [since $(\forall z \leq m) \sim R_1(x,z)$].

   From (1) and (4) it follows that $n > m$. From (2) and (3) it follows that it is also true that $m > n$, which is impossible.

5. We are to show that the following two characterizations of the recursive inseparability of the disjoint pair of sets (A,B) are equivalent:

   C_1: There is no recursive superset of A disjoint from B.

   C_2: For any disjoint recursively enumerable supersets ω_i of A and ω_j of B, ω_i cannot be the complement of ω_j.

To show that C_1 implies C_2, assume C_1. Now suppose ω_i, ω_j are disjoint supersets of A and B respectively. If ω_i were the complement of ω_j then ω_i would be a recursive superset of A disjoint from B, contrary to C_1. Therefore ω_i is not the complement of ω_j, which proves C_2.

Conversely, suppose C_2 holds. Let A' be any superset of A disjoint from B. We are to show that A' cannot be recursive and hence C_1 holds. Well, if A' were a recursive set ω_i, its complement $\overline{\omega_i}$ would be a recursively enumerable superset ω_j of B, disjoint from A, which would be contrary to C_2. Thus A' is not recursive.

6. We are given that $k(x, y)$ is a recursive function such that for all i and j,

(1) $k(i, j) \in \omega_i - \omega_j$ implies $k(i, j) \in A_2$;

(2) $k(i, j) \in \omega_j - \omega_i$ implies $k(i, j) \in A_1$.

Now consider any recursively enumerable superset ω_i of A_1 and any recursively enumerable superset ω_j of A_2. We are to show that $k(i, j) \in \omega_i$ iff $k(i, j) \in \omega_j$ [and thus that (A_1, A_2) is completely effectively inseparable under $k(x, y)$].

If $k(i, j) \in \omega_i - \omega_j$, then $k(i, j) \in A_2$ [by (1)], and hence $k(i, j) \in \omega_j$ [since $A_2 \subseteq \omega_j$], but no number in $\omega_i - \omega_j$ could possibly be in ω_j, and therefore $k(i, j) \notin \omega_i - \omega_j$. By a symmetric argument, $k(i, j) \notin \omega_j - \omega_i$. Consequently, $k(i, j) \in \omega_i$ iff $k(i, j) \in \omega_j$.

This completes the proof.

7. Since $\delta(i, j) \in \omega_j$ iff $\delta(i, j) \in A_1$ and $\delta(i, j) \in \omega_i$ iff $\delta(i, j) \in A_2$, then $\delta(i, j) \in \omega_j - \omega_i$ iff $\delta(i, j) \in A_1 - A_2$ and $\omega_i - \omega_j$ iff $\delta(i, j) \in A_2 - A_1$. Thus $\delta(x, y)$ is a Kleene function for $(A_1 - A_2, A_2 - A_1)$. By the separation principle, there are recursively enumerable sets K_1 and K_2 such that $A_1 - A_2 \subseteq K_1$ and $A_2 - A_1 \subseteq K_2$. Since $\delta(x, y)$ is a Kleene function for $(A_1 - A_2, A_2 - A_1)$, and $A_1 - A_2 \subseteq K_1$ and $A_2 - A_1 \subseteq K_2$, it follows that $\delta(x, y)$ is a Kleene function for the recursively enumerable pair (K_1, K_2). [In general, it is easily verified that if $\zeta(x, y)$ is a Kleene function for a pair (B_1, B_2), and if B_1', B_2' are disjoint supersets of $B_1 B_2$ respectively, then $\zeta(x, y)$ must be a Kleene function for (B_1', B_2').]

8. As mentioned in the hint: Let $f(x)$ be the function that semi-reduces (A_1, A_2) to (B_1, B_2), and let $t(y)$ be a recursive function given by the iteration theorem such that for all x and y, $x \in \omega_{t(y)}$ iff $f(x) \in \omega_y$ [thus $\omega_{t(y)} = \{x : f(x) \in \omega_y\}$]. Also $f(x) \in \omega_y$ iff $x \in f^{-1}(\omega_y)$ and so $x \in \omega_{t(y)}$

iff $x \in f^{-1}(\omega_y)$. Let k be a Kleene function for the (disjoint) Kleene pair $(A_1 A_2)$. We now let $k'(x, y)$ be the function $f(k(t(x), t(y)))$, and we will show that $k'(x, y)$ is a Kleene function for the pair (B_1, B_2). [At this point, the reader might like to try showing this without reading further.]

First we show that if $k'(i, j) \in \omega_j - \omega_i$, then $k'(i, j) \in B_1$.

Well, suppose $k'(i, j) \in \omega_j - \omega_i$. Thus $f(k(t(i), t(j))) \in \omega_j - \omega_i$, and so

$$k(t(i), t(j)) \in f^{-1}(\omega_j - \omega_i).$$

But

$$f^{-1}(\omega_j - \omega_i) = f^{-1}(\omega_j) - f^{-1}(\omega_i) = \omega_{t(j)} - \omega_{t(i)}$$

[since $f^{-1}(\omega_j) = \omega_{t(j)}$ and $f^{-1}(\omega_i) = \omega_{t(i)}$, as we have shown]. Thus

$$k(t(i), t(j)) \in \omega_{t(j)} - \omega_{t(i)},$$

so that $k(t(i), t(j)) \in A_1$ [since $k(x, y)$ is a Kleene function for (A_1, A_2)], and therefore $f(k(t(i), t(j))) \in B_1$ [since (A_1, A_2) is semi-reducible to (B_1, B_2) under $f(x)$]. Thus $k'(i, j) \in B_1$.

This proves that if $k'(i, j) \in \omega_j - \omega_i$, then $k'(i, j) \in B_1$.

By a similar argument, if $k'(i, j) \in \omega_i - \omega_j$, then $k'(i, j) \in B_2$. Thus $k'(x, y)$ is a Kleene function for (B_1, B_2).

9. If (A, B) is semi-doubly universal, then by definition the Kleene pair (K_1, K_2) is semi-reducible to (A, B). Hence [by Proposition 2] (A, B) is in turn a Kleene pair, and therefore completely effectively inseparable [by Proposition 1].

10. Suppose $\varphi(x, y)$ is a doubly generative function for (A, B). Let c be any index of the empty set, and thus for every number i, the set ω_i is disjoint from ω_c. Thus $\varphi(x, c)$ [as a function of x] is a generative function for A, because for every number i, ω_i is disjoint from ω_c, so that $\varphi(i, c) \in A$ iff $\varphi(i, c) \in \omega_i$ [by (1)]. Similarly, $\varphi(c, x)$ [as a function of x] is a doubly generative function for the set B.

11. The set $\{x : x \in \omega_n \lor x \in A\}$ is simply $\omega_n \cup A$, and the set $\{x : x \in \omega_n \lor x \in B\}$ is simply $\omega_n \cup B$. Using the t_1 and t_2 recommended in the hint, we have:

(1) $\omega_{t_1(n)} = \omega_n \cup A$;

(2) $\omega_{t_2(n)} = \omega_n \cup B$.

Thus A and ω_n are both subsets of $\omega_{t_1(n)}$ for all n, and B and ω_n are both subsets of $\omega_{t_2(n)}$ for all n. Also as in the hint, we take

$$\varphi(x, y) = \zeta(t_1(y), t_2(x)).$$

Given two numbers i and j, let $k = \varphi(i,j)$. Thus $k = \zeta(t_1(j), t_2(i))$. Since (A, B) is completely effectively inseparable under $\zeta(x, y)$ and $A \subseteq \omega_{t_1(j)}$ and $B \subseteq \omega_{t_2(i)}$ [remember, $A \subseteq \omega_{t_1(n)}$ and $B \subseteq \omega_{t_2(n)}$ for all n], and $\zeta(x, y)$ is a completely effective inseparability function for (A, B), then

(a) $k \in \omega_{t_1(j)}$ iff $k \in \omega_{t_2(i)}$.

Now, consider any disjoint pair (ω_i, ω_j). We are to show that $k \in A$ iff $k \in \omega_i$ and $k \in B$ iff $k \in \omega_j$, and thus that (A, B) is doubly generative under $\varphi(x, y)$.

Suppose that $k \in A$. Then $k \in \omega_{t_1(j)}$ [since $A \subseteq \omega_{t_1(n)}$ for every n]. Hence $k \in \omega_{t_2(i)}$ [by (a)]. Thus $k \in \omega_i \cup B$ [since $\omega_{t_2(i)} = \omega_i \cup B$]. But $k \notin B$, since B is disjoint from A. Thus $k \in \omega_i$. This proves that if $k \in A$ then $k \in \omega_i$.

Conversely, suppose that $k \in \omega_i$. Then $k \in \omega_{t_2(i)}$ [since $\omega_i \subseteq \omega_{t_2(i)}$]. Thus $k \in \omega_{t_1(j)}$ [by (a)], and since $\omega_{t_1(j)} = \omega_j \cup A$, then $k \in \omega_j \cup A$. But since ω_j is disjoint from ω_i we have the $k \notin \omega_j$, and so $k \in A$. Thus if $k \in \omega_i$, then $k \in A$.

This proves that $k \in A$ iff $k \in \omega_i$. The proof that $k \in B$ iff $k \in \omega_j$ is similar.

12. We are given that (A_1, A_2) is doubly generative under $g(x, y)$.

   (a) Suppose that $\omega_i = \mathbb{N}$ and $\omega_j = \emptyset$. Of course ω_i is disjoint from ω_j, so that $g(i, j) \in \mathbb{N}$ iff $g(i, j) \in A_1$ [and also $g(i, j) \in \emptyset$ iff $g(i, j) \in A_2$], but since $g(i, j) \in \mathbb{N}$, it follows that $g(i, j) \in A_1$.
   (b) Similarly, if $\omega_i = \emptyset$ and $\omega_j = \mathbb{N}$, then $g(i, j) \in A_2$.
   (c) If $\omega_i = \emptyset$ and $\omega_j = \emptyset$, since $\emptyset$ is obviously disjoint from $\emptyset$, then $g(i, j) \in A_1$ iff $g(i, j) \in \emptyset$, and $g(i, j) \in \emptyset$ iff $g(i, j) \in A_2$. But since $g(i, j) \notin \emptyset$, then $g(i, j) \notin A_1$ and $g(i, j) \notin A_2$.

13. By Problem 16 of the last chapter, for any recursively enumerable set B there is a recursive function $t(y)$ such that for all numbers i, if $i \in B$ then $\omega_{t(i)} = \mathbb{N}$, and if $i \notin B$, then $\omega_{t(i)} = \emptyset$.

   Now, let (B_1, B_2) be any disjoint pair of recursively enumerable sets. There are recursive functions $t_1(i), t_2(i)$ such that for any number i,

   (1) If $i \in B_1$, then $\omega_{t_1(i)} = \mathbb{N}$;
   (2) If $i \notin B_1$, then $\omega_{t_1(i)} = \emptyset$;

(3) If $i \in B_2$, then $\omega_{t_2(i)} = \mathbb{N}$;

(4) If $i \notin B_2$, then $\omega_{t_2(i)} = \emptyset$.

Now assume that conditions C_1 and C_2 hold between the function $g(x, y)$ and the pair (A_1, A_2).

(a) Suppose $i \in B_1$. Then $\omega_{t_1(i)} = \mathbb{N}$ [by (1)]. Since B_2 is disjoint from B_1, then $i \notin B_2$, and hence $\omega_{t_2(i)} = \emptyset$ [by (4)], and thus the pair $(\omega_{t_1(i)}, \omega_{t_2(i)})$ is $(\mathbb{N}, \emptyset)$, and so $g(t_1(i), t_2(i)) \in A_1$. This proves that $i \in B_1$ implies $g(t_1(i), t_2(i)) \in A_1$.

(b) The proof that $i \in B_2$ implies $g(t_1(i), t_2(i)) \in A_2$ is similar. Thus (B_1, B_2) is semi-reducible to (A_1, A_2) under the function $g(t_1(x), t_2(x))$.

(c) Suppose that condition C_3 also holds. Now suppose that $i \notin B_1$ and $i \notin B_2$. Then by (2) and (4), $\omega_{t_1(i)} = \emptyset$ and $\omega_{t_2(i)} = \emptyset$, and therefore $g(t_1(i), t_2(i)) \notin A_1$ and $g(t_1(i), t_2(i)) \notin A_2$ [by C_3]. Thus $g(t_1(x), t_2(x))$ reduces (B_1, B_2) to (A_1, A_2).

14. We are given that (A, B) is doubly co-productive under the recursive function $g(x, y)$.

D_0: This is obvious!

D_1: Suppose $\omega_i = \{g(i, j)\}$ and $\omega_j = \emptyset$. We are to show that $g(i, j) \in A$. Consider the following conditions:

(1) ω_i is disjoint from ω_j;

(2) ω_j is disjoint from B;

(3) ω_i is disjoint from A.

If these three conditions all hold, then $g(i, j)$ would be outside all four sets A, B, ω_i, ω_j [by the definition of double co-productivity]. Hence $g(i, j)$ would be outside ω_i, hence outside $\{g(i, j)\}$, which is not possible. Therefore it cannot be that all three of these conditions hold, but (1) and (2) do hold, therefore (3) doesn't hold, which means that ω_i is not disjoint from A. Thus $g(i, j)$ is not disjoint from A, so that $g(i, j) \in A$ This proves D_1.

D_2: The proof of D_2 is symmetric to that of D_1.

15. By Theorem 3 there are recursive functions $t_1(i, j)$ and $t_2(i, j)$ such that for all numbers i and j,

(1) $\omega_{t_1(i,j)} = \{x : M_1(x, i, t_1(i, j), t_2(i, j))\}$;

(2) $\omega_{t_2(i,j)} = \{x : M_2(x, j, t_1(i, j), t_2(i, j))\}$.

Thus,

$(1')$ $\omega_{t_1(i,j)} = \{x : x \in \omega_i \wedge x = g(t_1(i,j), t_2(i,j))\}$
$= \omega_i \cap \{g(t_1(i,j), t_2(i,j))\};$

$(2')$ $\omega_{t_2(i,j)} = \{x : x \in \omega_j \wedge x = g(t_1(i,j), t_2(i,j))\}$
$= \omega_j \cap \{g(t_1(i,j), t_2(i,j))\}.$

16. Consider any disjoint pair (ω_i, ω_j). Let $\varphi(i,j) = g(t_1(i,j), t_2(i,j))$, where $t_1(y_1, y_2), t_2(y_1, y_2)$ are as given by the lemma. Then by the lemma,

(a) $\omega_{t_1(i,j)} = \omega_i \cap \{\varphi(i,j)\};$
(b) $\omega_{t_2(i,j)} = \omega_j \cap \{\varphi(i,j)\}.$

We first wish to show:

(1) If $\varphi(i,j) \in \omega_i$, then $\varphi(i,j) \in A;$
(2) If $\varphi(i,j) \in \omega_j$, then $\varphi(i,j) \in B.$

To show (1), suppose that $\varphi(i,j) \in \omega_i$. Then $\omega_i \cap \{\varphi(i,j)\} = \{\varphi(i,j)\}$. Hence, by (a), $\omega_{t_1(i,j)}$ is the set $\{\varphi(i,j)\}$.

Since $\varphi(i,j) \in \omega_i$ and ω_i is disjoint from ω_j, then $\varphi(i,j) \notin \omega_j$, and therefore $\omega_j \cap \{\varphi(i,j)\} = \emptyset$. Hence, by (b), $\omega_{t_2(i,j)} = \emptyset$. Thus $\omega_{t_1(i,j)}$ is the unit set $\{\varphi(i,j)\}$ and $\omega_{t_2(i,j)}$ is the empty set $\emptyset$, and so by condition D_1 of the definition of "weakly doubly co-productive," $g(t_1(i,j), t_2(i,j)) \in A$. Thus $\varphi(i,j) \in A$, which proves (1).

The proof of (2) is similar (using D_2 instead of D_1).

It remains to prove the converses of (1) and (2). To this end, we first establish:

(3) If $\varphi(i,j) \in A \cup B$, then $\varphi(i,j) \in \omega_i \cup \omega_j.$

We will prove the equivalent propositions that if $\varphi(i,j) \notin \omega_i \cup \omega_j$, then $\varphi(i,j) \notin A \cup B.$

Well, suppose $\varphi(i,j) \notin \omega_i \cup \omega_j$. Then $\varphi(i,j) \notin \omega_i$ and $\varphi(i,j) \notin \omega_j$. Since $\varphi(i,j) \notin \omega_i$, then $\omega_i \cap \{\varphi(i,j)\} = \emptyset$. Then by (a), $\omega_{t_1(i,j)} = \emptyset$. Similarly, since $\varphi(i,j) \notin \omega_j$, then $\omega_{t_2(i,j)} = \emptyset$. Hence by condition D_0 of the definition of "weakly doubly productive,"

$$g(t_1(i,j), t_2(i,j)) \notin A \cup B.$$

Thus $\varphi(i,j) \notin A \cup B$. This proves that if $\varphi(i,j) \notin \omega_i \cup \omega_j$, then $\varphi(i,j) \notin A \cup B$, and therefore if $\varphi(i,j) \in A \cup B$, then $\varphi(i,j) \in \omega_i \cup \omega_j$, which is (3).

Now, to prove the converse of (1), suppose $\varphi(i,j) \in A$. Then $\varphi(i,j) \in A \cup B$, so that $\varphi(i,j) \in \omega_i \cup \omega_j$. Since $\varphi(i,j) \in A$, then $\varphi(i,j) \notin B$ [because B is disjoint from A]. Therefore $\varphi(i,j) \notin \omega_j$ [because if $\varphi(i,j) \in \omega_j$, then by (2) $\varphi(i,j)$ would be in B, which it isn't]. Since $\varphi(i,j) \in \omega_i \cup \omega_j$ but $\varphi(i,j) \notin \omega_j$, then $\varphi(i,j) \in \omega_i$. Thus if $\varphi(i,j) \in A$, then $\varphi(i,j) \in \omega_i$, which is the converse of (1). The proof of the converse of (2) is similar.

17. (a) Suppose that A is recursively enumerable. Let $M(x,y)$ be the recursively enumerable relation $x \in \omega_y \vee x \in A$. By the iteration theorem, there is a recursive function $t(i)$ such that for all i, $\omega_{t(i)} = \{x : M(x,i)\}$. Thus

$$\omega_{t(i)} = \{x : x \in \omega_i \vee x \in A\} = \omega_i \cup A.$$

(b) Suppose (A,B) is effectively inseparable under $\zeta(x,y)$ and A and B are both recursively enumerable. Now let (ω_i, ω_j) be a disjoint pair such that ω_i is disjoint from A and ω_j is disjoint from B. Since ω_i is disjoint from both ω_j and A, then ω_i is disjoint from $\omega_j \cup A$. Since B is disjoint from both ω_j and A, then B is disjoint from $\omega_j \cup A$. Thus ω_i and B are both disjoint from $\omega_j \cup A$, so that $\omega_i \cup B$ is disjoint from $\omega_j \cup A$.

Thus, $\omega_j \cup A$ and $\omega_i \cup B$ are disjoint supersets of A and B respectively. By (a), there are recursive functions t_1 and t_2 such that $\omega_{t_1(j)} = \omega_j \cup A$ and $\omega_{t_2(i)} = \omega_i \cup B$. Therefore $\omega_{t_1(j)}, \omega_{t_2(i)}$ are disjoint supersets of A and B respectively. Because the pair (A,B) is assumed to be effectively inseparable under $\zeta(x,y)$, $\zeta(t_1(j), t_2(i))$ [which we define $g(i,j)$ to be] lies outside both $\omega_{t_1(j)}$ and $\omega_{t_2(i)}$. But $\omega_{t_1(j)} = \omega_j \cup A$ and $\omega_{t_2(i)} = \omega_i \cup B$, which means $g(i,j)$ must of course lie outside both ω_i and ω_j (and A and B as well). Thus we see that (A,B) is doubly generative under the function $g(x,y)$.

Chapter 7

Metamathematical Applications

In *The Beginner's Guide* [Smullyan, 2014] we studied the system of Peano Arithmetic. Some of the results regarding that system, as well as those regarding other formal systems, depend on the use of the logical connectives and quantifiers, but not all do. Those that do not can be neatly generalized to more abstract systems called *representation systems*, which I introduced in the *Theory of Formal Systems* [Smullyan, 1961]. In this chapter, I generalize further using systems I call *simple systems*, in terms of which it is shown that many results of Peano Arithmetic and other systems are essentially results of recursion theory in a new dress.

In what follows, *number* shall mean natural number, and *function* shall mean function whose arguments and values are natural numbers. Any argument of a function ranges over the whole set of natural numbers.

I. Simple Systems

By a simple system S I shall mean a pair (P, R) of disjoint sets, together with a function $neg(x)$ and a function $r(x, y)$ satisfying condition C_1 below.

For any numbers h and n, we abbreviate $neg(h)$ by h', and $r(h, n)$ by $h(n)$, or sometimes by $r_h(n)$.

C_1: For any numbers h and n, $h'(n) \in P$ iff $h(n) \in R$, and $h'(n) \in R$ iff $h(n) \in P$.

Intended Applications

With any first-order system $\mathcal{F}$ of arithmetic we associate a simple system $\mathcal{S}$ as follows: We arrange all the sentences of $\mathcal{F}$ in an infinite sequence $S_1, S_2, \ldots, S_n, \ldots$ according to the magnitude of their Gödel numbers. We refer to n as the *index* of the sentence S_n. We take one particular variable x and arrange all formulas with x as their only free variable in an infinite sequence $\varphi_1(x), \varphi_2(x), \ldots, \varphi_n(x), \ldots$ according to the magnitude of their Gödel numbers. We refer to n as the *index* of $\varphi_n(x)$. We arrange all formulas with the two variables x and y as their only free variables in an infinite sequence $\psi_1(x,y), \psi_2(x,y), \ldots, \psi_n(x,y), \ldots$, according to the magnitudes of their Gödel numbers, and for any numbers n, x and y, we define $s(n,x,y)$ to be the index of the sentence $\psi_n(\bar{x}, \bar{y})$.

For the *associated* simple system $\mathcal{S}$ of $\mathcal{F}$, we take P to be the set of indices of the provable sentences, and take R to be the set of indices of the refutable sentences. For any numbers h and n, we take $r(h,n)$ to be the index of the sentence $\varphi_h(\bar{n})$ [i.e. the sentence resulting from substituting the number $\bar{n}$ (the name of n) for all free occurrences of x in $\varphi_h(x)$]. We take h' [which is the abbreviation of $neg(h)$] to be the index of the formula $\sim\varphi_h(x)$ Of course $\varphi_{h'}(x)$ is provable (refutable) iff $\varphi_h(x)$ is refutable (provable). And so the items (P, R), $neg(h)$, and $r(x,y)$ constitute a simple system. Consequently, whatever we prove for simple systems in general will hold for Peano Arithmetic and other systems.

Returning to simple systems in general, in accordance with intended applications, we might call members of P *provable numbers*, and members of R *refutable numbers*. We will call a number *decidable* if it is in either P or R, and call it *undecidable* otherwise. We call $\mathcal{S}$ *complete* if every number is decidable, otherwise *incomplete*.

We call two numbers *equivalent* if whenever either one is in P, so is the other, and if either one is in R, so is the other.

We will refer to the function $r(x,y)$ as the *representation function for sets* of the system $\mathcal{S}$, and we say that a number h *represents* the set A of all numbers n such that $h(n) \in P$. Thus, to say that h represents A is to say that for all n, $h(n) \in P$ iff $n \in A$. We call A *representable* in $\mathcal{S}$ if some number h represents it.

We say that h *completely represents* A iff h represents A and h' represents the complement $\overline{A}$ of A.

We say that h *defines* A if, for all numbers n:
(1) $n \in A$ implies $h(n) \in P$;
(2) $n \in \overline{A}$ implies $h(n) \in R$.

We say that A is *definable* (*completely representable*) if some number h defines (completely represents) A.

We say that h *contra-represents* A if for all n the number $h(n) \in R$ iff $n \in A$.

Problem 1. (a) Show that h defines A iff h completely represents A. (b) Show that h contra-represents A iff h' represents A, so that A is contra-representable iff it is representable.

For any function $f(x)$ and set A, by $f^{-1}(A)$ is meant the set of all n such that $f(n) \in A$. Thus $n \in f^{-1}(A)$ iff $f(n) \in A$.

Recall that $r_h(n)$ is $r(h, n)$ and that $r_h(n) = h(n)$, where $r(x, y)$ is the representation function for sets.

Problem 2. Show that h represents A iff $A = r_h^{-1}(P)$. Show also that h contra-represents A iff $A = r_h^{-1}(R)$.

Problem 3. Prove that if some set A is represented in $\mathcal{S}$ and its complement $\overline{A}$ is not, then $\mathcal{S}$ is incomplete.

For any set A, by A^* we shall mean the set of all numbers h such that $h(h) \in A$.

Problem 4. Prove that neither the complement $\overline{P^*}$ of P^*, nor the complement $\overline{R^*}$ of R^*, is representable in $\mathcal{S}$.

By the *diagonalization* of a number h we shall mean the number $h(h)$. By the *skew diagonalization* of h we shall mean the number $h(h')$.

It follow from Problems 3 and 4 that if either P^* or R^* is representable in $\mathcal{S}$, then $\mathcal{S}$ is incomplete. The following problem yields a sharper result.

Problem 5. Prove the following:
(a) If h represents R^*, then its diagonalization $h(h)$ is undecidable.

(b) If h represents P^*, then its skew-diagonalization $h(h')$ is undecidable.

The following problem yields still stronger results.

Problem 6. Prove the following:
(a) If h represents some superset of R^* disjoint from P^*, then $h(h)$ is undecidable.
(b) If h represents some superset of P^* disjoint from R^*, then $h(h')$ is undecidable.

Why are these results stronger than those of Problem 5?

Discussion

Gödel's original incompleteness proof essentially involved representing the set P^*. [In the *Theory of Formal Systems* [Smullyan, 1961], I pointed out that incompleteness also follows if R^* is representable.] To represent P^* Gödel had to assume that the system was ω-consistent. It was John Barkley Rosser [1936] who was able to eliminate the hypothesis of ω-consistency by representing, not R^*, but some superset of R^* disjoint from P^*.

The Diagonal Function $d(x)$

We define $d(x)$ to be $r(x, x)$, which is $x(x)$. We note that for any set A the set A^* is $d^{-1}(A)$.

Problem 7. Prove that for any sets A and B:
(a) If $A \subseteq B$, then $f^{-1}(A) \subseteq f^{-1}(B)$.
(b) If A is disjoint from B, then $f^{-1}(A)$ is disjoint from $f^{-1}(B)$.
(c) $f^{-1}(\overline{A})$ is the complement of $f^{-1}(A)$.

In particular, taking $d(x)$ for $f(x)$,
(a_1) If $A \subseteq B$, then $A^* \subseteq B^*$.
(a_2) If A is disjoint from B, then A^* is disjoint from B^*.
(a_3) $\overline{A}^* = \overline{A^*}$.

Admissible Functions

We shall call a function $f(x)$ *admissible* if for every number h there is a number k such that for every n the number $k(n)$ is equivalent to $h(f(n))$.

Separation

Given a disjoint pair (A, B), we say that a number h *separates* A from B (in S) if h represents a superset of A and contra-represents some superset of B.

We say that h *exactly separates* A from B if h both represents A itself and contra-represents B.

The following problem is key.

Problem 8. Suppose $f(x)$ is admissible. Prove the following:
(a) If A is representable, so is $f^{-1}(A)$. If A is contra-representable, so is $f^{-1}(A)$.
(b) If a disjoint pair (A, B) is exactly separable in S, then so is the pair $(f^{-1}(A), f^{-1}(B))$.

Normal Systems

We shall call S a *normal system* if the diagonal function $d(x)$ is admissible. This means that for every number h there is a number — which we shall denote $h^{\#}$ — such that for all n the numbers $h^{\#}(n)$ is equivalent to $h(d(n))$.

Problem 9. Suppose S is normal. If A is representable in S, does it necessarily follow that A^{*} is representable in S?

Problem 10. Show that for any normal system, neither $\overline{P}$ nor $\overline{R}$ is representable.

Fixed Points

By a *fixed point* of h is meant a number n such that $h(n)$ is equivalent to n.

Problem 11. Prove that if S is normal, then every number h has a fixed point.

Problem 12. Prove the following:
(a) If h represents R, then any fixed point of h is undecidable.
(b) If h represents P, then any fixed point of h' is undecidable.

Exercise. Show the following strange facts:

(1) If h represents some superset of R disjoint from P, then $h(h)$ is undecidable.
(2) If h represents some superset of P disjoint from R, then $h(h')$ is undecidable.

We will later have important use for the following:

Problem 13. Show that no superset A of R^*, with A disjoint from P^*, can be definable in S. [**Hint:** Show that if $R^* \subseteq A$, and h defines A in S, then h is in both A and P^* (and hence that A is not disjoint from P^*).]

Exercise. Show that if $P^* \subseteq A$, and h defines A in S, then h' is in both A and R^* (and hence that A is not disjoint from R^*).

II. Standard Simple Systems

Now recursion theory enters the picture. We shall call a simple system S a standard simple system if the sets P and R are recursively enumerable, and the functions $neg(x)$, $r(x, y)$ and $s(x, y, z)$ are recursive. In what follows, S will be assumed to be a standard simple system. [Recall that the function $s(n, x, y)$ is the index of the sentence $\psi_n(\bar{x}, \bar{y})$.]

Problem 14. Show that for any set A,
(1) If A is recursively enumerable, so is A^*.
(2) If A is recursive, so is A^*.
More generally, show that for any recursive function $f(x)$,
(1') If A is recursively enumerable, so is $f^{-1}(A)$.
(2') If A is recursive, so is $f^{-1}(A)$.

Problem 15. Show that every set representable in S is recursively enumerable, and that every set definable in S is recursive.

Problem 16. Show the following:
(a) If some recursively enumerable but non-recursive set is representable in S, then S is incomplete.
(b) If all recursively enumerable sets are representable in S, then S is incomplete.

Question: Suppose we are given, not that all recursively enumerable sets are representable in S, but only that all *recursive* sets are representable in S. Is that enough to guarantee that S is incomplete? The reader might like to try answering this on his or her own. The answer will appear in this text later on.

Undecidability and Incompleteness

The word "undecidable" has two very different meanings in the literature, which might sometimes cause confusions. On the one hand, a sentence of a system is called *undecidable* in the system if it is neither provable nor refutable in the system. On the other hand, the system as a whole is called *undecidable* if the set of provable sentence is not recursively solvable (or: under an effective Gödel numbering, the set of Gödel numbers of the provable sentences is not recursive).

We shall thus call a simple system S *undecidable* if the set P is not recursive.

The two senses of *undecidable*, though different, have the following important link:

Theorem 1. *If S is undecidable (but standard), then S is incomplete (some number is undecidable in S, i.e. outside both P and R).*

Problem 17. Prove Theorem 1.

It was known prior to 1961 that if all recursively enumerable sets are representable in a system, then the system is undecidable. In the *Theory of Formal Systems* [Smullyan, 1961] I proved the stronger fact that if all *recursive* sets are representable in a system, the system is undecidable. My proof goes over to simple systems.

Theorem 2. *If all recursive sets are representable in S, then S is undecidable.*

Problem 18. Prove Theorem 2. [**Hint:** Use the fact already proved that if P is recursive, so is P^* (Problem 14).]

Remark. For a standard system S, if all recursive sets are representable in S, then S is undecidable by Theorem 2, hence incomplete by Theorem 1.

Thus if all recursive sets are representable in a standard system S, then S is incomplete, which answers the question raised following Problem 16.

Recursive Inseparability

Problem 19. If the pair (P, R) is recursively inseparable, does it necessarily follow that S is undecidable?

Problem 20. Show that for any recursive function $f(x)$ and any disjoint pair (A, B), if the pair $(f^{-1}(A), f^{-1}(B))$ is recursively inseparable, then the pair (A, B) is also recursively inseparable. Conclude that if (P^*, R^*) is recursively inseparable, so is (P, R).

We next wish to prove:

Theorem 3. *If every recursive set is definable in S then the pair (P^*, R^*) is recursively inseparable, and so is the pair (P, R).*

Problem 21. Prove Theorem 3. [**Hint:** Use Problem 13.]

Recall that, given a disjoint pair (A, B), we say that a number h *separates* A from B (in S) if h represents a superset of A and contra-represents some superset of B. And we say that h *exactly separates* A from B if h both represents A and contra-represents B.

We next wish to prove the following theorem, which has the interesting corollaries that follow it:

Theorem 4. *If some disjoint pair (A_1, A_2) is recursively inseparable in S, but separable in S, then the pair (P, R) is recursively inseparable.*

Corollary 1. *If some recursively inseparable pair is separable in S, then S is undecidable.*

Corollary 2. *If some recursively inseparable pair is separable in S, then S is incomplete (assuming S is standard).*

Note: Corollary 2 is essentially Kleene's *symmetric* form of Gödel's theorem [Kleene, 1952]. Kleene proved this for a particular pair of recursively inseparable sets, but the proof works for any recursively inseparable pair of recursively enumerable sets.

Problem 22. Prove Theorem 4.

Rosser Systems

We shall call S a *Rosser system* if for any two recursively enumerable sets A and B, there is a number h that separates $A - B$ from $B - A$. [If A and B are disjoint, then of course such an h also separates A from B.]

We shall call a Rosser system S an *exact Rosser system*, if for any two disjoint recursively enumerable sets A and B, there is a number h that *exactly* separates A from B.

Gödel's proof of the incompleteness of systems like Peano Arithmetic involved representing a certain recursively enumerable set, but the only way of representing recursively enumerable sets that was known at that time involved the hypothesis of ω-consistency. However, Andrzej Ehrenfeucht and Solomon Feferman [1960] found a way of representing all recursively enumerable sets in Peano Arithmetic, without having to assume ω-consistency! In terms of simple systems, what they showed was that if S is a Rosser system in which all recursive functions of one argument are admissible, then all recursively enumerable sets are representable in S. Later came a stronger result, which, for simple systems, is:

Theorem 5. [After Putnam, Smullyan, 1961] *If S is a Rosser system in which all recursive functions of one argument are admissible, then S is an exact Rosser system.*

Theorem 5 is an obvious consequence of the following two propositions:

Proposition 1. *If S is a Rosser system, then some doubly universal pair is exactly separable in S.*

Proposition 2. *If some doubly universal pair is exactly separable in S and if all recursive functions of one argument are admissible, then every disjoint pair of recursively enumerable sets is exactly separable in S.*

Problem 23. Prove Propositions 1 and 2. [**Hint**: For Proposition 1, show that if the Kleene pair (K_1, K_2) (or any other effectively inseparable pair of recursively enumerable sets) is separable in S, then some doubly universal pair is exactly separable in S.]

Effective Rosser Systems

We shall call S an *effective Rosser system* if there is a recursive function $\Pi(x, y)$ such that for any numbers i and j, $\Pi(i, j)$ separates $(\omega_i - \omega_j)$ from $(\omega_j - \omega_i)$. Such a function $\Pi(x, y)$ will be called a *Rosser function*.

Effective Rosser systems have some very interesting properties, as we will see. For one thing, we will soon see that every effective Rosser system is an exact Rosser system (a result that is incomparable in strength with Theorem 5).

The following proposition is key for this, as well as for some further results.

Proposition 3. *If S is an effective Rosser system, then for any recursively enumerable relations $R_1(x, y), R_2(x, y)$, there is a number h such that for any n:*
(1) $R_1(n, h) \wedge \sim R_2(n, h)$ *implies* $h(n) \in P$;
(2) $R_2(n, h) \wedge \sim R_1(n, h)$ *implies* $h(n) \in R$.

Problem 24. Prove Proposition 3. [**Hint:** Apply the weak double recursion theorem to the relations $R_1(x, \Pi(y, z)), R_2(x, \Pi(y, z))$, where $\Pi(x, y)$ is a Rosser function for S.]

Now we can prove:

Theorem 6. *If S is an effective Rosser system, then S is an exact Rosser system.*

Problem 25. Prove Theorem 6. [**Hint:** Let $R_1(x, y)$ be the relation $x \in A \vee y(x) \in R$, and let $R_2(x, y)$ be the relation $x \in B \vee y(x) \in P$. Recall that $y(x) = r(y, x)$. Apply Proposition 3 to $R_1(x, y)$ and $R_2(x, y)$.]

Rosser Fixed-Point Property

We say that S has the *Rosser fixed point property* if for any two recursive functions $f_1(x)$ and $f_2(x)$, there is a number h such that for all n for which $\omega_{f_1(n)}$ is disjoint from $\omega_{f_2(n)}$:
(1) $h(n) \in \omega_{f_1(n)}$ implies $h(n) \in P$;
(2) $h(n) \in \omega_{f_2(n)}$ implies $h(n) \in R$.

Theorem 7. *Every effective Rosser system has the Rosser fixed point property.*

Problem 26. Prove Theorem 7. [**Hint:** Given recursive functions $f_1(x)$ and $f_2(x)$, apply Proposition 3 to the relations $y(x) \in \omega_{f_1(x)}$ and $y(x) \in \omega_{f_2(x)}$.]

Uniform Incompleteness

By the *principal part* of a simple system $\mathcal{S}$ we mean the pair (P, R).

We call a simple system $\mathcal{S}'$ an *extension* of a simple system $\mathcal{S}$ if $P \subseteq P'$ and $R \subseteq R'$, where (P', R') is the principal part of $\mathcal{S}'$, and the representation function $r(x, y)$ and the function $neg\,(x)$ are the same in both $\mathcal{S}$ and $\mathcal{S}'$.

Now consider an infinite sequence $\mathcal{S} = \mathcal{S}_1, \mathcal{S}_2, \ldots, \mathcal{S}_n, \ldots$, where each $\mathcal{S}_{n+1}$ is an extension of $\mathcal{S}_n$. Such a sequence is called an *effective sequence*, or a *recursively enumerable sequence*, if the relations $x \in P_y$ and $x \in R_y$ are both recursively enumerable. The system $\mathcal{S}$ is called *uniformly incompletable* if, for any recursively enumerable sequence $\mathcal{S}_1$, $\mathcal{S}_2, \ldots, \mathcal{S}_n, \ldots$, of extensions of $\mathcal{S}$, there is a number h such that, for every n, the number $h(n)$ is an undecidable number of $\mathcal{S}_n$ (i.e. $h(n)$ is outside $P_n \cup R_n$).

The following is one of the main results.

Theorem 8. *Every effective Rosser system is uniformly incompletable.*

Problem 27. Prove Theorem 8. [**Hint:** Given a recursively enumerable sequence of extensions of $\mathcal{S}$, show that there are recursive functions $f_1(x)$ and $f_2(x)$ such that for all n the set $\omega_{f_1(n)}$ is R_n and $\omega_{f_2(n)}$ is P_n. Then use the Rosser fixed point property.]

$$* * *$$

Before leaving this chapter, I should say a little more about the relation of simple systems to the first-order arithmetical systems in standard formalization of Tarski [1953].

Consider such a first-order system $\mathcal{F}$ and its associated simple system $\mathcal{S}$, as defined much earlier in this chapter.

$\mathcal{F}$ is called a *Rosser system*, more specifically a Rosser system for sets, iff its associated simple system $\mathcal{S}$ is a Rosser system, as defined here, which is equivalent to the condition that for any pair of recursively enumerable sets A and B there is a formula $\varphi_h(x)$ such that for all n, if $n \in A - B$, then $\varphi_h(\bar{n})$ is provable (in $\mathcal{F}$), and if $n \in B - A$, then $\varphi_h(\bar{n})$ is refutable in $\mathcal{F}$. [In terms of the associated simple system $\mathcal{S}$, if $n \in A - B$, then $h(n) \in P$ and if $n \in B - A$, then $h(n) \in R$.]

The system $\mathcal{F}$ is called an *exact Rosser system* (*for sets*) if for any pair of disjoint recursively enumerable sets A and B, there is a predicate H such that for all n, $n \in A$ iff $H(n)$ is provable, and $n \in B$ iff $H(n)$ is refutable.

In the literature, a function $f(x)$ is said to be *definable* if there is a formula $\psi(x, y)$ such that for all n and m:

(1) If $f(n) = m$, then $\psi(\bar{n}, \bar{m})$ is provable;

(2) If $f(n) \neq m$, then $\psi(\bar{n}, \bar{m})$ is refutable;

(3) The sentence $\forall x \forall y ((\psi(\bar{n}, x) \wedge \psi(\bar{n}, y)) \supset x = y)$ is provable.

As far as I know, the term *admissible function* does not appear in the literature, except in my own writings. However, if a formula $F(x, y)$ defines a function $f(x)$, then for any formula $H(x)$, if we take $G(x)$ to be the formula $\exists y (F(x, y) \wedge H(x))$, then it can be verified that for every n, not only is it the case that $G(\bar{n})$ is equivalent to $H(\overline{f(n)})$, but the very sentence $G(\bar{n}) \equiv H(\overline{f(n)})$ is provable in the system. For details, see Theorem 2, Chapter 8 of *Gödel's Incompleteness Theorems* [Smullyan, 1992], in which I use the term "strongly definable" for what I am calling "definable" here.

Thus every definable function $f(x)$ is indeed admissible. The Ehrenfeucht–Feferman Theorem, as stated, is that any consistent Rosser system in which all recursive functions $f(x)$ are *definable* is an exact Rosser system.

In the context of a first-order system $\mathcal{F}$, a formula $\psi(x, y)$ is said to *separate* a relation $R_1(x, y)$ from a relation $R_2(x, y)$ if for all n and m,

(1) $R_1(n, m)$ implies that $\psi(\bar{n}, \bar{m})$ is provable.

(2) $R_2(n, m)$ implies that $\psi(\bar{n}, \bar{m})$ is refutable.

$\mathcal{F}$ is called a *Rosser system for binary relations* if for any two recursively enumerable relations $R_1(x, y)$, $R_2(x, y)$, there is a formula $\psi(x, y)$ that separates the relation $R_1(x, y) \wedge {\sim} R_2(x, y)$ from $R_2(x, y) \wedge {\sim} R_1(x, y)$. Shepherdson's Theorem is that every consistent Rosser system for binary relations is an exact Rosser system for sets. This appears to be incomparable in strength with the Putnam–Smullyan Theorem, which is that every consistent Rosser system in which all recursive functions of one argument are definable is an exact Rosser system for sets.

The proof of Shepherdson's Theorem (which can be found in [Shepherdson, 1961]) is not very different from my proof that every effective Rosser system is an exact Rosser system. Indeed, this theorem and its proof borrowed heavily from Shepherdson's Theorem and proof.

Solutions to the Problems of Chapter 7

1. (a) It is obvious that if h completely represents A, then h defines A. To show the converse, suppose h defines A. Then for all n,

    (1) $n \in A$ implies $h(n) \in P$;
    (2) $n \in \overline{A}$ implies $h(n) \in R$.

    We are to show the converses of (1) and (2).

    Suppose $h(n) \in P$. Then $h(n) \notin R$ (since R is disjoint from P). Thus by (2) $n \in \overline{A}$ doesn't hold, so that $n \in A$. The proof that $h(n) \in R$ implies that $n \in \overline{A}$ is similar.

   (b) h contra-represents A iff for every n, $n \in A$ holds iff $h \in R$, which is true iff $h'(n) \in P$, which is true iff h' represents A. Conversely, suppose h' represents A. Then for all $n, n \in A$ iff $h'(n) \in P$, which is true iff $h(n) \in R$. And so h contra-represents A.

2. We are to show that h represents A iff $A = r_h^{-1}(P)$. We first recall that $r_h(n) = h(n)$, since $r_h(n) = r(h, n) = h(n)$.

    (1) Suppose h represents A. Then $n \in A$ iff $h(n) \in P$, which is true iff $r_h(n) \in P$, which is true iff $n \in r_h^{-1}(P)$. Thus $n \in A$ iff $n \in r_h^{-1}(P)$. Since this is true for all n, $A = r_h^{-1}(P)$.

    ' (2) Conversely suppose $A = r_h^{-1}(P)$. Then $n \in A$ iff $n \in r_h^{-1}(P)$. But $n \in r_h^{-1}(P)$ is true iff $r_h(n) \in P$, which is true iff $h(n) \in P$. Thus $n \in A$ iff $h(n) \in P$. Since this is true for all n, h represents A.

    The proof that h contra-represents A iff $A = r_h^{-1}(R)$ is similar.

3. We prove the equivalent proposition that if $\mathcal{S}$ is complete, then the complement of any representable set is representable.

    Well, suppose $\mathcal{S}$ is complete, and that h represents A. We will show that h' represents $\overline{A}$. Consider any number n.

    (1) Suppose $n \in \overline{A}$. Then $n \notin A$. Hence $h(n) \notin P$ [since h represents A]. Hence $h(n) \in R$ [since $\mathcal{S}$ is complete]. Thus $h'(n) \in P$. Thus $n \in \overline{A}$ implies $h'(n) \in P$. .

(2) Conversely, suppose $h'(n) \in P$. Then [by condition C_1] $h(n) \in R$, so that $h(n) \notin P$. Thus $n \notin A$ [since h represents A], so that $n \in \overline{A}$. Thus $h'(n) \in P$ implies that $n \in \overline{A}$.

By (1) and (2), h' represents $\overline{A}$.

4. If some number h represents $\overline{P^*}$, we can reach contradiction as follows: $h(h) \in P$ iff $h \in \overline{P^*}$ [since h represents $\overline{P^*}$, for all n we have $n \in \overline{P^*}$ iff $h(n) \in P$]. But $h \in \overline{P^*}$ iff $h \notin P^*$, which is true iff $h(h) \notin P$. Thus we get the contradiction that $h(n) \in P$ iff $h(n) \notin P$, so that no number h represents $\overline{P^*}$.

By a similar argument (using R instead of P), the set $\overline{R^*}$ cannot be contra-representable. Thus $\overline{R^*}$ cannot be representable [by Problem 1(b)].

5. (a) Suppose h represents R^*. Then, for all n, $h(n) \in P$ iff $n \in R^*$, which is true iff $n(n) \in R$. Taking h for n, we have $h(h) \in P$ iff $h(h) \in R$. Since $h(h)$ cannot be in both P and R, $h(h)$ must be outside both of them, and is thus undecidable.

(b) Suppose h represents P^*. Then, for all n, $h(n) \in P$ iff $n \in P^*$. Thus $h(h') \in P$ iff $h' \in P^*$, and $h' \in P^*$ iff $h'(h') \in P$, which is true iff $h(h') \in R$. Thus $h(h') \in P$ iff $h(h') \in R$, and since P is disjoint from R, $h(h')$ is outside both P and R, and is thus undecidable.

6. (a) Suppose h represents some superset A of R^* that is disjoint from P^*. We must show that $h(h)$ is undecidable.

Well, if $h \in P^*$, then $h(h) \in P$, and $h \in A$ [since h represents A], contrary to the fact that P^* is disjoint from A. Thus $h \notin P^*$, since assuming so led to a contradiction, and so $h(h) \notin P$.

If $h \in A$, then $h(h) \in P$ [since h represents A]. Thus $h \in P^*$, again contrary to the fact that A is disjoint from P^*. Thus $h \notin A$, since assuming so led to a contradiction. Then $h \notin R^*$ [since $R^* \subseteq A$], and so $h(h) \notin R$.

Thus $h(h) \notin P$ and $h(h) \notin R$ so that $h(h)$ is undecidable.

(b) Suppose h represents some superset A of P^* disjoint from R^*. We must show that $h(h')$ is undecidable.

Well, if $h' \in A$, then $h(h') \in P$ [since h represents A]. Hence $h'(h') \in R$ [by condition C_1], and so $h' \in R^*$, contrary to the fact that R^* is disjoint from A. Therefore $h' \notin A$, since assuming so led to a contradiction. Then $h' \notin P^*$ [since $P^* \subseteq A$], and so $h'(h') \notin P$, and therefore $h(h') \notin R$.

If $h' \in R^*$, then $h'(h') \in R$, so that $h(h') \in P$, so that $h' \in A$ [which is represented by h], which is again contrary to the fact that R^* is disjoint from A. Hence $h' \notin R^*$, since assuming so led to a contradiction. Then $h'(h') \notin R$, and therefore $h(h') \notin P$.

Thus $h(h') \notin R$ and $h(h') \notin P$, and therefore $h(h')$ is undecidable.

The reason why these results are stronger than those of Problem 5 is this: To begin with, since P is disjoint from R, then P^* is disjoint from R^* (why?). Therefore, if h represents R^*, it automatically represents some superset of R^* disjoint from P^*, namely R^* itself. [Any set is both a subset and a superset of itself.] Similarly, if h represents P^*, then it represents a superset of P^* disjoint from R^*, namely R^* itself. Thus the results of Problem 5 are but special cases of those of Problem 6.

7. (a) Suppose $A \subseteq B$. For any n, if $n \in f^{-1}(A)$, then $f(n) \in A$. Thus $f(n) \in B$ [since $A \subseteq B$], and hence $n \in f^{-1}(B)$. Thus $n \in f^{-1}(A)$ implies $n \in f^{-1}(B)$, and so $f^{-1}(A) \subseteq f^{-1}(B)$.

(b) Suppose A is disjoint from B. If $n \in f^{-1}(A)$, then $f(n) \in A$. Thus $f(n) \notin B$. Thus $n \in f^{-1}(A)$ implies that $n \notin f^{-1}(B)$, and thus $f^{-1}(A)$ is disjoint from $f^{-1}(B)$.

(c) $n \in f^{-1}(\overline{A})$ iff $f(n) \in \overline{A}$, which is true iff $f(n) \notin A$, which is true iff $n \notin f^{-1}(A)$, which is true iff $n \in \overline{f^{-1}(A)}$. Thus $n \in f^{-1}(\overline{A})$ iff $n \in \overline{f^{-1}(A)}$, and since this holds for every n, we have

$$f^{-1}(\overline{A}) = \overline{f^{-1}(A)}.$$

The statements (a_1), (a_2), and (a_3) clearly follow from statements (a), (b), and (c) when $f = d(x)$, because $d^{-1}(A)$ is the set A^* for any set A.

8. Suppose $f(x)$ is admissible. Given h, let k be such that $k(n)$ is equivalent to $h(f(n))$ for all n.

(a) We assert that if h represents a set A, then k represents $f^{-1}(A)$, and that if h contra-represents A, then k contra-represents $f^{-1}(A)$.

To prove the former, suppose h represents A. Then $n \in f^{-1}(A)$ iff $f(n) \in A$, which is true iff $h(f(n)) \in P$ [since h represents A], which is true iff $k(n) \in P$. Thus, for all $n, n \in f^{-1}(A)$ iff $k(n) \in P$, and thus k represents $f^{-1}(A)$.

The proof that if h contra-represents A, then k contra-represents $f^{-1}(A)$ is similar (just replace "P" by "R" and "represents" by "contra-represents").

(b) Suppose h exactly separates the set A from the set B. Then h represents A and contra-represents B. Then, by the solution to part (a), k represents $f^{-1}(A)$ and contra-represents $f^{-1}(B)$, and so k exactly separates $f^{-1}(A)$ from $f^{-1}(B)$.

9. Of course it follows! If S is normal, the diagonal function $d(x)$ is admissible. Hence if A is representable, so is $d^{-1}(A)$ [by Problem 8], and $d^{-1}(A)$ is the set A^*.

10. For a normal system, if $\overline{P}$ were representable, then $\overline{P}^*$ would be representable [by Problem 9]. But $\overline{P}^* = \overline{P^*}$ [by Problem 7], so that $\overline{P^*}$ would be representable, which is contrary to Problem 4. A similar argument show that $\overline{R}$ is not representable.

11. Suppose S is normal. For any numbers h and n, $h^{\#}(n)$ is equivalent to $h(d(n))$, so that $h^{\#}(h^{\#})$ is equivalent to $h(d(h^{\#}))$. But $d(h^{\#}) = h^{\#}(h^{\#})$ and so $h^{\#}(h^{\#})$ is equivalent to $h(h^{\#}(h^{\#}))$, which means that $h^{\#}(h^{\#})$ is a fixed point of h.

12. (a) Suppose h represents R and that n is some fixed point of h [i.e. n is equivalent to $h(n)$]. Since h represents R, then $n \in R$ iff $h(n) \in P$. But $h(n) \in P$ iff $n \in P$ [since n is a fixed point of h]. Thus $n \in R$ iff $n \in P$. But R and P are assumed to be disjoint, so the fixed point n of h can belong to neither P nor R, which means that n must be undecidable.

(b) Suppose h represents P and that n is some fixed point of h'. Since h represents P, then $n \in P$ iff $h(n) \in P$. But $h(n) \in P$ iff $h'(n) \in R$ [by condition C_1]. And $h'(n) \in R$ iff $n \in R$ [since n is a fixed point of h']. Thus $n \in P$ iff $n \in R$, and so again the arbitrary fixed point n of h' is undecidable.

13. Suppose $R^* \subseteq A$ and h defines A in S. We will show that A cannot be disjoint from P^*. Suppose $h \notin A$. Then $h(h) \in R$ [since h defines A]. Hence $h \in R^*$, which would mean $h \in A$ [since $R^* \subseteq A$], which is impossible. Thus it cannot be that $h \notin A$, which implies $h \in A$. It then further follows that $h(h) \in P$ [since h defines, and hence represents, A]. Therefore $h \in P^*$. Thus $h \in A$ and $h \in P^*$, so that A is not disjoint from P^*.

14. We will first consider the general case.

(1') Let $f(x)$ be a recursive function and suppose the set A is recursively enumerable. For any number x, $x \in f^{-1}(A)$ iff $f(x) \in A$

which is true iff $\exists y (y = f(x) \wedge y \in A)$. Thus $f^{-1}(A)$ is recursively enumerable.

(2′) If A is also recursive, then A and its complement $\overline{A}$ are both recursively enumerable, so that $f^{-1}(A)$ and $f^{-1}(\overline{A})$ are both recursively enumerable, as just shown. However, $f^{-1}(\overline{A})$ is the complement of $f^{-1}(A)$, so that $f^{-1}(\overline{A})$ is recursive.

To apply the general case to (1) and (2): Since the function $r(x, y)$ is recursive, so is the diagonal function $d(x)$ [which is $r(x, x)$], and since $A^* = d^{-1}(A)$, it follows from (1′) and (2′) that if A is recursively enumerable (recursive), so is A^*.

15. We must show that (a) every set representable in $\mathcal{S}$ is recursively enumerable, and that (b) every set definable in $\mathcal{S}$ is recursive.

   (a) Suppose h represents A. Then for all n, $n \in A$ iff $r(h, n) \in P$. Let $f(x) = r(h, x)$ Since the function $r(x, y)$ is recursive, so is the function $f(x)$. Thus for all, $n \in A$ iff $f(n) \in P$, and so $A = f^{-1}(P)$. Since P is recursively enumerable, and $f(x)$ is a recursive function, then $f^{-1}(P)$ is recursively enumerable [by Problem 14], and thus A is recursively enumerable.

   (b) If A is completely representable [which is equivalent to A being definable, by Problem 1(a)], then A and $\overline{A}$ are both representable, hence both recursively enumerable, and so A is then recursive.

16. (a) Suppose that A is a recursively enumerable set that is not recursive, and that A is representable in $\mathcal{S}$. Since A is not recursive, its complement $\overline{A}$ is not recursively enumerable, hence not representable [since only recursively enumerable sets are representable in $\mathcal{S}$, by Problem 15]. Thus A is a representable set whose complement is not representable. Hence $\mathcal{S}$ is incomplete [by Problem 3].

   (b) This follows from (a) and the fact that there exists a recursively enumerable set that is not recursive, such as Post's complete set K [if all recursively enumerable sets were representable in $\mathcal{S}$, K would also have to be recursively enumerable].

17. Suppose $\mathcal{S}$ is undecidable and effective. Then the set P is not recursive, and so its complement $\overline{P}$ is not recursively enumerable. Yet the set R is recursively enumerable [since $\mathcal{S}$ is effective], and so R is not the complement of P. But R is disjoint from P, which means that some n must be outside both P and R, and is thus undecidable. Thus $\mathcal{S}$ is incomplete.

18. Suppose all recursive sets are representable in S. If P were recursive, then P^* would be recursive [by Problem 14]. Thus $\overline{P^*}$ would be recursive, hence representable in S, which would be contrary to Problem 4. Therefore P is not recursive, and S is thus undecidable [by the definition of an undecidable system].

19. Of course it does! If a disjoint pair of sets (A, B) is recursively inseparable, then by definition no recursive superset of A can be disjoint from B. Thus A cannot be recursive [for it were, then A itself would be a recursive superset of A disjoint from B]. Similarly, B cannot be recursive.

   Thus, if (P, R) were recursively inseparable, then P would not be recursive, so that S would be undecidable.

20. We are to show that if $f(x)$ is a recursive function and (A, B) is any disjoint pair, if $(f^{-1}(A), f^{-1}(B))$ is disjoint and recursively inseparable, so is (A, B). We will prove the equivalent proposition that if a disjoint pair (A, B) is recursively separable, so is $(f^{-1}(A), f^{-1}(B))$.

   Well, suppose that A is disjoint from B and that A is recursively separable from B. Then there is a recursive superset A' of A disjoint from B. Hence $f^{-1}(A)$ is disjoint from $f^{-1}(B)$ [by Problem 7(b)]. And $f^{-1}(A)$ is recursive [by Problem 14]. Thus $f^{-1}(A')$ is a recursive superset of $f^{-1}(A)$ that is disjoint from $f^{-1}(B)$, and so $f^{-1}(A)$ is recursively separable from $f^{-1}(B)$.

   Since $(P^*, R^*) = (d^{-1}(P)d^{-1}(R))$ and $d(x)$ is a recursive function, it follows from what was just shown that if (P^*R^*) is recursively inseparable, so is (P, R).

21. Suppose all recursive sets are definable in S. We will show that (P^*, R^*) is recursively inseparable, from which we will be able to conclude that so is the pair (P, R) [by Problem 20].

   Let A be any superset of R^* disjoint from P^*. By Problem 13, A cannot be definable in S. If A were recursive, then it *would* be definable in S [since the problem assumes this]. Thus A cannot be recursive. Consequently there is no recursive superset of R^* disjoint from P^*, which means that (P^*, R^*) is recursively inseparable.

22. Suppose (A_1, A_2) is disjoint and recursively inseparable, and that h separates A_1 from A_2 in S. Then h represents some superset B_1 of

A_1 and contra-represents some superset B_2 of A_2. In this situation B_1 and B_2 must be disjoint [because $n \in B_1$ iff $h(n) \in P$ and $n \in B_2$ iff $h(n) \in R$, but P and R are disjoint]. Since (A_1, A_2) is recursively inseparable, and $A_1 \subseteq B_1$ and $A_2 \subseteq B_2$, then the pair (B_1, B_2) is recursively inseparable (verify this!). Now, by Problem 2, $B_1 = r_h^{-1}(P)$ and $B_2 = r_h^{-1}(R)$, where $r(x, y)$ is the representation function for sets. Thus the pair $(r_h^{-1}(P), r_h^{-1}(R))$ is recursively inseparable, and since the function $r_h(x)$ is recursive [because $r(x, y)$ is], then the pair (P, R) is recursively inseparable [by Problem 20].

Now, Corollary 1 follows because if (P, R) is recursively inseparable, then of course P is not recursive, so that S is undecidable.

Corollary 2 then follows from Corollary 1 and Theorem 1 (if S is effective but undecidable, then S is incomplete).

23. (a) To prove Proposition 1, suppose K_1 is separable from K_2 in S and let h effect the separation. Then h represents some superset A_1 of K_1 and contra-represents some superset A_2 of K_2. Since P is disjoint from R, then A_1 is disjoint from A_2. Of course, h exactly separates A_1 from A_2.

   Since (K_1, K_2) is effectively inseparable, so is the pair (A_1, A_2) [verify this!]. Since S is an effective system, then A_1 and A_2 are recursively enumerable sets. Thus (A_1, A_2) is an effectively inseparable pair of recursively enumerable sets, hence is doubly universal [by Theorem 11 of Chapter 6]. Thus h exactly separates the doubly universal pair (A_1, A_2).

  (b) To prove Proposition 2, suppose that some doubly universal pair (U_1, U_2) is exactly separable in S and that all recursive functions of one argument are admissible in S. Let (A, B) be a disjoint pair of recursively enumerable sets that we wish to exactly separate in S.

   Since (U_1, U_2) is doubly universal, then there is a recursive function $f(x)$ such that $A = f^{-1}(U_1)$ and $B = f^{-1}(U_2)$. By hypothesis, $f(x)$ is admissible (all recursive functions of one argument are). Thus, by Problem 8, $f^{-1}(U_1)$ is exactly separable from $f^{-1}(U_2)$, and thus A is exactly separable from B.

24. Suppose $\Pi(x, y)$ is a Rosser function for S. Given recursively enumerable relations $R_1(x, y), R_2(x, y)$, by the weak double recursion theorem (Theorem 1 of Chapter 6), taking $M_1(x, y, z)$ to be $R_1(x, \Pi(y, z))$

and $M_2(x,y,z)$ to be $R_2(x, \Pi(y,z))$, there are numbers a and b such that:

$$\omega_a = \{x : R_1(x, \Pi(a,b))\};$$
$$\omega_b = \{x : M_2(x, \Pi(a,b))\}.$$

We let h be the number $\Pi(a,b)$. Then $\omega_a = \{x : R_1(x,h)\}$ and $\omega_b = \{x : R_2(x,h)\}$. Thus $\omega_a - \omega_b = \{x : R_1(x,h) \wedge \sim R_2(x,h)\}$ and $\omega_b - \omega_a = \{x : R_2(x,h) \wedge \sim R_1(x,h)\}$. Since $\Pi(x,y)$ is a Rosser function, then h, which is $\Pi(a,b)$, separates $\omega_a - \omega_b$ from $\omega_b - \omega_a$, and thus separates $\{x : R_1(x,h) \wedge \sim R_2(x,h)\}$ from $\{x : R_2(x,h) \wedge \sim R_1(x,h)\}$.

25. Let A and B be disjoint recursively enumerable sets. Let $R_1(x,y)$ be the relation $x \in A \vee y(x) \in R$ and let $R_2(x,y)$ be the relation $x \in B \vee y(x) \in P$. Applying Proposition 3 to the pair (R_1, R_2), there is a number h such that for all n,

$R_1(n,h) \wedge \sim R_2(n,h)$ implies $h(n) \in P$ and
$R_2(n,h) \wedge \sim R_1(n,h)$ implies $h(n) \in R$. Thus

(1) $[(n \in A \vee h(n) \in R) \wedge \sim(n \in B \vee h(n) \in P)] \supset h(n) \in P$;
(2) $[(n \in B \vee h(n) \in P) \wedge \sim(n \in A \vee h(n) \in R)] \supset h(n) \in R$.

Also,

(3) $\sim(n \in A \wedge n \in B)$ [since A is disjoint from B];
(4) $\sim(h(n) \in P \wedge h(n) \in R)$ [since P is disjoint from R].

From (1), (2), (3) and (4) it follows by propositional logic that $n \in A$ iff $h(n) \in P$, and $n \in B$ iff $h(n) \in R$.

To see this, let us use the following abbreviations:

$$p_1 \text{ for } n \in A; \quad p_2 \text{ for } n \in B;$$
$$q_1 \text{ for } h(n) \in P; \quad q_2 \text{ for } h(n) \in R.$$

Thus we have:
(1) $[(p_1 \vee q_2) \wedge \sim(p_2 \vee q_1)] \supset q_1$;
(2) $[(p_2 \vee q_1) \wedge \sim(p_1 \vee q_2)] \supset q_2$;
(3) $\sim(p_1 \wedge p_2)$;
(4) $\sim(q_1 \wedge q_2)$.

We are to infer that $p_1 \equiv q_1$ and $p_2 \equiv q_2$.

(a) First we show that $p_1 \supset q_1$. Well, suppose p_1. Then $p_1 \vee q_2$ is true. Thus (1) is equivalent to

$$(1') \sim(p_2 \vee q_1) \supset q_1.$$

Since p_1 is true (by assumption), then p_2 is false [by (3)], so that $p_2 \vee q_1$ is equivalent to q_1, so that $(1')$ is equivalent to

$$(2') \sim q_1 \supset q_1.$$

From this q_1 follows, which proves that $p_1 \supset q_1$.

(b) For the converse, suppose q_1 is true. Then $p_2 \vee q_1$ is true, and so (2) is equivalent

$$(2') \sim (p_1 \vee q_2) \supset q_2.$$

Since q_1 is true (by assumption), then q_2 is false [by (4)], so that $p_1 \vee q_2$ is equivalent to p_1, so that $(2')$ is equivalent to

$$(2'') \sim p_1 \supset q_2.$$

Since q_2 is false, it follows from $(2'')$ that p_1 is true. This proves that $q_1 \supset p_1$. Thus $p_1 \equiv q_1$.

The proof that $p_2 \equiv q_2$ is similar (just interchange p_1 with p_2 and q_1 with q_2).

Note: This proof is a modification of a proof of a result of John Shepherdson.

26. Suppose S is an effective Rosser system. Given two recursive functions $f_1(x)$ and $f_2(x)$ let $R_1(x, y)$ be the relation $y(x) \in \omega_{f_1(x)}$ and let $R_2(x, y)$ be the relation $y(x) \in \omega_{f_2(x)}$.

Applying Proposition 3 to these two relations, there is a number h such that for every n,

(1) $h(n) \in \omega_{f_1(n)} \wedge \sim(h(n) \in \omega_{f_2(n)})$ implies $h(n) \in P$;

(2) $h(n) \in \omega_{f_2(n)} \wedge \sim(h(n) \in \omega_{f_1(n)})$ implies $h(n) \in R$.

If $\omega_{f_1(n)}$ is disjoint from $\omega_{f_2(n)}$, then (1) and (2) are respectively equivalent to:

$(1')$ $h(n) \in \omega_{f_1(n)}$ implies $h(n) \in P$;

$(2')$ $h(n) \in \omega_{f_2(n)}$ implies $h(n) \in R$.

Thus S has the Rosser fixed point property.

27. Consider a recursively enumerable sequence $S = S_1, S_2, \ldots, S_n, \ldots$, of extensions of S. The relation $x \in R_y$ is recursively enumerable. Hence by the iteration theorem, there is a recursive function $f_1(x)$ such that for all n, we have $\omega_{f_1(n)} = \{x : x \in R_n\}$. Thus $\omega_{f_1(n)} = R_n$. Similarly, there is a recursive function $f_2(x)$ such that, for all n, we have $\omega_{f_1(n)} = P_n$.

Now suppose that S is effectively a Rosser system. Then S has the Rosser fixed point property, since for every n the sets $\omega_{f_1(x)}$ and $\omega_{f_2(x)}$

are disjoint. So by the Rosser fixed point property, there is a number h such that for all n:

(1) $h(n) \in \omega_{f_1(n)}$ implies $h(n) \in P$;

(2) $h(n) \in \omega_{f_2(n)}$ implies $h(n) \in R$.

Thus,

(1') $h(n) \in R_n$ implies $h(n) \in P$;

(2') $h(n) \in P_n$ implies $h(n) \in R$.

Since $P \subseteq P_n$ and $R \subseteq R_n$, then by (1') and (2'), we have:

(1') $h(n) \in R_n$ implies $h(n) \in P_n$;

(2') $h(n) \in P_n$ implies $h(n) \in R_n$.

Since P_n is disjoint from R_n, it must be the case that $h(n) \notin P_n$ and $h(n) \notin R_n$. Thus $\mathcal{S}$ is uniformly incompletable.

Part III
Elements of Combinatory Logic

Chapter 8

Beginning Combinatory Logic

The subject of combinatory logic, initiated by Moses Schönfinkel [1924], is a most elegant one, as the reader will see. It plays a significant role in computer science and provides yet another approach to recursion theory. In this book we provide only the beginnings of the subject, hoping to whet the reader's appetite for more.

By an *applicative system* is meant a set $\mathcal{C}$ together with an operation that assigns to each element x and each element y (in that order) an element denoted (xy). In combinatory logic we often omit parentheses, with the understanding that parentheses are to be restored to the left: xyz abbreviates $((xy)z)$, not $(x(yz))$. Also $xyzw$ abbreviates $(((xy)z)w)$, etc. Note that this is not the same as the common set of rules for eliminating parentheses in school mathematics and even in many college mathematics courses, so it takes some getting used to. But have courage, for it will become natural fairly quickly. (But always remember to put outside parentheses around an expression when you replace a variable by a compound expression; for instance, replacing the variable x with SI in the expression $S(Kx)K$ would result in $S(K(SI))K$.) Also note that, in general, in applicative systems xy is not the same as yx, nor is $x(yz)$ the same as $(xy)z$ (i.e. the commutative and associative rules of elementary arithmetic do not generally apply here).

The elements of $\mathcal{C}$ are called *combinators*. It is to be understood that $\mathcal{C}$ contains more than one element.

Fixed Points

An element y is called a *fixed point* of an element x if $xy = y$. Fixed points play a key role in combinatory theory.

The Composition Condition

For any elements A, B and C, we say that C *combines* A with B if $Cx = A(Bx)$ for every element x. We say that *the composition condition holds* if for every A and B, there is some C that combines A with B.

Introducing the Combinator M, the Mocker (the Mockingbird)

A combinator M will be called a *mocker*, or a *duplicator*, if

$$Mx = xx$$

for every element x. Here is our first interesting result:

Theorem 1. *The following two conditions are sufficient to ensure that every element A has a fixed point:*
C_1: *The composition condition holds.*
C_2: *There is a mocker M.*

Problem 1. Prove Theorem 1.

   Note: Theorem 1 states a basic fact of combinatorial logic. The solution, though quite brief, is very ingenious, and derives ultimately from the work of Gödel. It is related to many of the fixed point results in several areas in mathematical logic.

   The solution given in the proof of Theorem 1 reveals more information than is given in the statement of the theorem. This extra information is important, and we record it as follows:

Theorem 1*. *If C combines x with M, then CC is a fixed point of x.*

Egocentric Elements

We shall call an element x *egocentric* if it is a fixed point of itself, i.e. if $xx = x$.

Problem 2. Prove that if conditions C_1 and C_2 of the hypothesis of Theorem 1 hold, then at least one element is egocentric.

Agreeable Elements

We shall say that two elements A and B *agree* on an element x if $Ax = Bx$. We shall call an element A *agreeable* if for every element B, there is at least one element x on which A and B agree. Thus A is agreeable if, for every B, $Ax = Bx$ for at least one x.

Problem 3. (a) [A Variant of Theorem 1] Suppose we are given condition C_1 of the hypothesis of Theorem 1 [i.e. for every A and B, there is some C that combines A with B], but instead of being given that there is a mocker M, we are given that at least one element A is agreeable. Prove that every element has a fixed point. (b) Theorem 1 is really but a special case of (a). Why?

Problem 4. As an exercise, show that if the composition condition C_1 holds, and if C combines A with B, then if C is agreeable, so is A.

Problem 5. Again, assuming the composition condition C_1, show that for any elements A, B and C, there is an element E such that $Ex = A(B(Cx))$ for every x.

Compatible Elements

We will say that an element A is *compatible* with B if there are elements x and y such that $Ax = y$ and $By = x$. When such a pair of elements (x, y) exists, it is called a *cross point* of the pair (A, B).

Problem 6. Prove that if conditions C_1 and C_2 of Theorem 1 hold, then any two elements A and B are compatible, i.e. any two elements have a cross point.

Smug Elements

We shall call an element *smug* if it is compatible with itself.

Problem 7. If an element has a fixed point, is it necessarily smug?

Problem 8. Prove that if the composition condition C_1 holds and if there is also at least one smug element, then there is at least one element that has a fixed point.

Fixation

We shall say that the element A is *fixated on* B if $Ax = B$ for every element x.

Problem 9. Is it possible for an element A to be fixated on more than one element?

Problem 10. Which, if either, of the following two statements are true?
(a) If y is a fixed point of x, then x is fixated on y.
(b) If x is fixated on y, then y is a fixed point of x.

Narcissistic Elements

We shall call an element A *narcissistic* if it is fixated on itself, i.e. if $Ax = A$ for every x.

Problem 11. Which, if either, of the following statements is true?
(a) Every narcissistic element is egocentric.
(b) Ever egocentric element is narcissistic.

Problem 12. Determine whether or not the following statement is true: If A is narcissistic, then for all elements x and y, $Ax = Ay$.

Problem 13. If A is narcissistic, does it necessarily follow that for all x and y, $Axy = A$?

Problem 14. Narcissism is contagious! If A is narcissistic, then so is Ax for every x. Prove this.

Introducing the Combinator K, the Kestrel

We shall call a combinator K a *cancellator*, or a *kestrel*, if

$$Kxy = x$$

for every x and y. This K plays a key role in combinatory logic, as we shall see.

Note: Why the alternative word "kestrel"? Well, in my book *To Mock a Mockingbird* [Smullyan, 1985], I used talking birds as combinators. If one calls out the name of a bird y to a bird x the bird x calls back the name of some bird, and this bird I call xy. Thus xy is the bird named by x upon hearing the name of y. I gave bird names to the standard combinators of the literature, usually names whose first letter is the first letter of the combinator in the literature. For example, I call the bird representing the standard combinator K by the name *kestrel*. The standard combinators B, C, W and T, which we will soon come across, I respectively called the *bluebird, cardinal, warbler* and *thrush*. I called the mocker M a *mockingbird*, because in my bird model, $Mx = xx$ means that M's response to x is the same as x's response to x.

To my surprise, I recently found out that many of my bird names have been adopted in the more recent combinatory literature! For example, there is a book by Reginald Braithwaite entitled *Kestrels, Quirky Birds and Hopeless Egocentricity* [2013]. In light of this, I feel justified in sometimes using bird names for various combinators.

In what follows, K is a *kestrel*.

Problem 15. Prove that any fixed point of K is egocentric.

Problem 16. Prove that if Kx is egocentric, then x is a fixed point of K.

Problem 17. Prove that if Kx is a fixed point of K, then x is a fixed point of K.

Problem 18. In general, it is not the case that if $Ax = Ay$, then x must necessarily be y, but it is the case if A is a kestrel. Prove that if $Kx = Ky$, x must necessarily be y.

The principle expressed in Problem 18 (that if $Kx = Ky$, x must necessarily be y) is important and will be called *the cancellation law*.

Problem 19. Prove that for all x, $Kx \neq K$.

Problem 20. (a) Prove that no kestrel can be egocentric. (b) Prove that any fixed point of a kestrel is narcissistic.

Next I want to show that if C contains a kestrel, then C contains infinitely many elements (recall that C always contains more than one element).

Problem 21. Consider the set $\{K, KK, KKK, \ldots\}$. Is this set infinite?

Problem 22. What about the set $\{K, KK, K(KK), K(K(KK)), \ldots\}$, i.e. the set $\{K_1, K_2, \ldots, K_n, \ldots\}$ where $K_1 = K$ and $K_{n+1} = K(K_n)$ for all n. Is this set infinite?

Introducing the Combinator L, the Lark

We shall call a combinator L a *lark* if for all x and y, the following condition holds:

$$Lxy = x(yy).$$

The lark was not a standard combinator in the literature prior to the publication of my *To Mock a Mockingbird*, but it is so now. Several papers have been written about it. The lark plays a key role in my treatment of combinatory logic.

One nice thing about the lark is that without assuming conditions C_1 and C_2, the mere existence of a lark ensures that every element has a fixed point.

Problem 23. Prove this.

Problem 24. Prove that no element can be both a lark and a kestrel.

Problem 25. Prove that if a lark L is narcissistic, then L must be a fixed point of every element x.

Problem 26. Prove that no kestrel can be a fixed point of a lark.

However, it is possible for a lark to be a fixed point of a kestrel.

Problem 27. Show that if a lark L is a fixed point of a kestrel, then L must be a fixed point of every element.

An interesting thing about larks is that the existence of a lark L, without any other information, is enough to guarantee the existence of an egocentric element. In fact, we can write down an expression using only the letter L and parentheses that names an egocentric element. The shortest expression I have been able to find uses twelve occurrences of the letter L. Is there a shorter one? I raised this as an open problem in *To Mock a Mockingbird*, and I don't know if this problem has been solved yet.

Problem 28. Given a lark L, prove that there is an egocentric element, and then write down the name of such an element, using only the letter L and parentheses.

Discussion

Given two expressions involving only L and parentheses, is there a purely mechanical procedure to decide whether or not they name the same element? I posed this as an open problem in *To Mock a Mockingbird*, and it has been solved affirmatively by at least three people — Richard Statman in his paper "The Word Problem for Smullyan's combinatorial lark is decidable [1989], and by M. Sprenger and M. Wymann-Böni in "How to decide the lark" [1993].

Introducing the Combinator I, the Identity Combinator

By an *identity combinator* is meant an element I satisfying the condition that $Ix = x$ for every x.

Despite its apparently lowly status, the identity operator I will later be seen to play an important role. Meanwhile, here are some trivial facts about I:

(1) I is agreeable if and only if every element has a fixed point.

(2) If every pair of elements is compatible, then I is agreeable.

(3) I, though egocentric, cannot be narcissistic.

Problem 29. Prove these three facts. [It is not necessary to assume the composition principle, but it is necessary to use the fact that $\mathcal{C}$ contains more than one element.]

Problem 30. Prove that if $\mathcal{C}$ contains a lark L and an identity element I, then $\mathcal{C}$ contains a mocker M.

Sages

An element θ will be called a *sage* if its existence not only guarantees that every element has a fixed point, but θ is also such that when applied to any element x, it yields a fixed point of x. Thus, θ is a sage, if, for every x, θx is a fixed point of x, i.e., $x(\theta x) = \theta x$ for every element x.

In the classic literature of combinatory logic, sages are called *fixed point combinators*. I introduced the word *sage* in my book *To Mock a Mockingbird*, and the term *sage* has subsequently become standard in the literature, so I will use the term sage here. Much study has been devoted to sages, and in a later chapter, I will show many ways of deriving a sage from various other combinators. For now, let me just state that if $\mathcal{C}$ contains a lark L and a mocker M, and if the composition law holds, them $\mathcal{C}$ contains a sage.

Problem 30. Prove this.

Note: It follows from the last problem that if $\mathcal{C}$ contains a lark L and an identity combinator I, and obeys the composition condition, the $\mathcal{C}$ contains a sage, because in that case $\mathcal{C}$ must also contain a mocker M by Problem 30.

Solutions to the Problems of Chapter 8

1. Let C be an element that combines A with M. Thus, for every element x, we have $Cx = A(Mx)$. But $Mx = xx$, and so $Cx = A(xx)$ for every x. Now take C for x, and we see that $CC = A(CC)$ which shows us that CC is a fixed point of A. [Clever, eh?]

2. From conditions C_1 and C_2, there is a mocker M. Moreover, every element has a fixed point [by Theorem 1]. Thus M itself has a fixed point E. Thus $ME = E$. Also $ME = EE$ [since M is a mocker], and so $E = EE$, which means that E is egocentric. Thus any fixed point of M is egocentric.

   Remark. Since E is egocentric and $ME = E$, it follows that ME is egocentric. Doesn't the world "ME" tell its own tale?

3. (a) Suppose A is agreeable. Given any element x, by condition C_1 there is some element C that combines x with A. Thus $Cy = x(Ay)$ for every element y. Since A is agreeable, then there is some y^* such that $Cy^* = Ay^*$, and hence it follows from $Cy^* = x(Ay^*)$ that $Ay^* = x(Ay^*)$, and so Ay^* is a fixed point of x.

   (b) Theorem 1 is a special case of (a), since a duplicator M is obviously agreeable: M agrees with every element x on some element, namely on x [since $Mx = xx$].

4. We are given that the composition condition C_1 holds and that C combines A with B and that C is agreeable. We are to show that A is agreeable. Well, consider any element D. We are to show that A agrees with D on some element. Well, let E be an element that combines D with B. Now, C agrees with E on some element F, since C is agreeable. We will show that A agrees with D on BF.

Since C agrees with E on F, then $CF = EF$. Also $EF = D(BF)$, since E combines D with B. Thus $CF = EF = D(BF)$, and so $CF = D(BF)$. But also $CF = A(BF)$, since C combines A with B, and so $A(BF) = D(BF)$. Thus A agrees with D on BF.

5. We assume that the composition condition C_1 holds. Given A, B and C, let D combine B with C. Thus for any x, $Dx = B(Cx)$. Now let E combine A with D. Then $Ex = A(Dx) = A(B(Cx))$.

6. For any elements A and B, we are given that there is some C that combines A with B. Thus for all y, we have $Cy = A(By)$. We are also given that there is a mocker M, and so C has some fixed point y. Thus $y = Cy$ and $Cy = A(By)$, yielding $y = A(By)$. Now take x to be By, so that $y = Ax$ and $x = By$. Thus $Ax = y$ and $By = x$.

7. Of course it is: Suppose x is a fixed point of A, so that $Ax = x$. Now take y to be x itself. Then $Ax = y$ and $Ay = x$, which means that A is smug.

8. We are given that the composition condition C_1 holds and that there is a smug element A. Thus there are elements x and y such that $Ax = y$ and $Ay = x$. Since $Ax = y$, we can substitute Ax for y in the equation $Ay = x$, yielding $A(Ax) = x$. Now let B be an element that combines A with itself. Thus $Bx = A(Ax)$, and since $A(Ax) = x$, we have $Bx = x$. Thus B has the fixed point x.

9. Of course not! If A is fixed on x and A is fixed on y, then for any element z we see that $Az = x$ and $Az = y$ so that $x = y$.

10. It is obviously (b) that is true. Suppose x is fixated on y. Then $xz = y$ for every z. Hence, taking y for z, we see that $xy = y$, so that y is a fixed point of x.

11. It is obviously (a) that is true. Suppose A is narcissistic. Then $Ax = A$ for every x, so of course $AA = A$.

12. As given, the statement is obviously true: If A is narcissistic, then $Ax = A$ and $Ay = A$. Thus Ax and Ay are both equal to A, so that $Ax = Ay$.

13. Yes, it does: If A is narcissistic, then $Ax = A$, so that $Axy = Ay$. But also $Ay = A$, yielding $Axy = A$.

14. Suppose A is narcissistic. Then by the last problem $Axy = A$. Thus $(Ax)y = A$ for every y, which means that Ax is narcissistic.

15. Suppose $Kx = x$ (i.e that x is a fixed point of K). Then $Kxx = xx$. But also $Kxx = x$ (since $Kxy = x$ for every y). Thus Kxx is

equal to both xx and to x, so that $xx = x$, which means that x is egocentric.

16. Suppose Kx is egocentric. Thus $Kx(Kx) = Kx$. Also $Kx(Kx) = x$ (since $Kxy = x$ for every y). Thus $Kx(Kx)$ is equal to both Kx and x, so that $Kx = x$.

17. Suppose Kx is a fixed point of K. Then Kx is egocentric by Problem 15. Hence x is a fixed point of K by Problem 16.

18. Suppose $Kx = Ky$. Now, $Kxz = x$ for every z. Since $Kx = Ky$, it follows that $Kyz = x$ for every z. But also $Kyz = y$. Hence $x = y$.

19. We will show that if $Kx = K$, then for any elements y and z, it must be the case that $y = z$, which is contrary to the assumed condition that C contains more than one element.

    Well, suppose $Kx = K$. Then for any y and z, $Kxy = Ky$ and $Kxz = Kz$. But $Kxy = x$ and $Kxz = x$, so that $Ky = Kz$. Then by the cancellation law (see Problem 18), $y = z$.

20. (a) This is immediate from the last problem: Since $Kx \neq K$ for all x, then $KK \neq K$, so that K is not egocentric.
    (b) Suppose $KA = A$ (i.e. A is a fixed point of K). Then $Ax = KAx$ for all x. But also $KAx = A$ for all x. Thus $Ax = A$ for all x, which means that A is narcissistic.

21. Let $U_1 = K$; $U_2 = KK$; $U_3 = KKK$, etc. We show that the set $\{U_1, U_2, U_3, \ldots, U_n, \ldots\}$ is far from infinite. It contains only K and KK. We see this as follows:

    If $U_n = K$, then $U_{n+2} = KKK$, which is K (because K is a kestrel). Thus if $U_n = K$, then $U_{n+2} = K$, and, since $U_1 = K$, then $U_n = K$ for every odd number n.

    Also, if $U_n = KK$ then $U_{n+1} = KKK$ which is K (again because K is a kestrel). Thus, if $U_n = KK$ then $U_{n+2} = KK$ and since $U_2 = KK$ it follows that $U_n = KK$ for every even n. Therefore we see that, for all n, $U_n = K$ or $U_n = KK$.

22. This is a very different story! Let $K_1 = K$, $K_2 = KK$; $K_3 = K(KK)$ [which is $K(K_2)$], etc. Thus for each n, we have $K_{n+1} = K(K_n)$. The set $\{K_1, K_2, \ldots, K_n, \ldots\}$ is indeed infinite, as we will see.

    We will show that for any numbers n and m, if $n \neq m$, then $K_n \neq K_m$. To show this, it suffices to show that if $n < m$, then

$K_n \neq K_m$ (because if $n \neq m$, then either $n < m$ or $m < n$). Thus it suffices to show that for any positive integers n and b, $K_n \neq K_{n+b}$.

By Problem 19, $K \neq K(K_b)$; in fact $K \neq Kx$ for any x. Since $K = K_1$ and $K(K_b) = K_{b+1}$, we have:

$$(1) \ K_1 \neq K_{1+b}.$$

Next, by the cancellation law (Problem 18), if $Kx = Ky$, then $x = y$. Equivalently, if $x \neq y$, then $Kx \neq Ky$. Thus if $K_n \neq K_m$, then $KK_n \neq KK_m$, and so we have:

$$(2) \ \text{If } K_n \neq K_m, \quad \text{then } K_{n+1} \neq K_{m+1}.$$

Now, by (1), we have $K_1 \neq K_{1+b}$. Hence by repeated applications of (2), we see that $K_2 \neq K_{2+b}$, $K_3 \neq K_{3+b}, \ldots, K_n \neq K_{n+b}$. This concludes the proof.

23. We consider a lark L. For any x and y, $(Lx)y = x(yy)$. Now, taking Lx for y, we see that $(Lx)(Lx) = x((Lx)(Lx))$. Thus $(Lx)(Lx)$ is a fixed point of x! Note that $(Lx)(Lx)$ is the same element as $Lx(Lx)$ by how we define the use of parentheses (or rather the lack thereof).

24. Consider a kestrel K. For any x and y, $Kxy = x$. Taking K for both x and y, we have:

$$(1) \ KKK = K.$$

If K were a lark, then $Kxy = x(yy)$ for all x and y, so that for $x = K$ and $y = K$, we would have:

$$(2) \ KKK = K(KK).$$

For (1) and (2) it would follow that $K = K(KK)$, violating the fact that for no x can $K = Kx$ (by Problem 19). Thus K cannot be a lark.

25. Suppose that a lark L is narcissistic. Then $Lxy = L$ for every x and y by the solution to Problem 13. Taking Lx for y, we obtain $Lx(Lx) = L$. But $Lx(Lx)$ is a fixed point of x [by the solution to Problem 23]. Thus L is a fixed point of x.

26. Suppose that $LK = K$ (i.e. that K is a fixed point of L). Then $LKK = KK$. But $LKK = K(KK)$ (since L is a lark), so that $K(KK) = KK$. Then by the cancellation law, taking KK for x and K for y, we see that $KK = K$, which means that K is egocentric, which is contrary to Problem 20. Therefore K cannot be a fixed point of L.

27. If the lark L is a fixed point of the kestrel K, then $KL = L$. So, for any x, $KLx = Lx$. But $KLx = L$, since K is a kestrel. So we have $Lx = L$ for all x, and L is narcissistic. Thus L is a fixed point of every element x by Problem 25.

28. We first show that if x is any fixed point of LL, then xx must be egocentric.

    Well, suppose that $LLx = x$. Now $LLx = L(xx)$ [since L is a lark]. Thus $x = L(xx)$. Therefore $xx = L(xx)x$, which is $xx(xx)$. Thus xx is egocentric.

    We can actually compute a fixed point of LL, in terms of just L: As we saw in the solution of Problem 23, for any x, a fixed point of x is $Lx(Lx)$. Taking LL for x, we see that a fixed point of LL is $L(LL)(L(LL))$, and therefore an egocentric element is $L(LL)(L(LL))$ repeated, which is:
$$L(LL)(L(LL))(L(LL)(L(LL))).$$

29. (1) Suppose I is agreeable. Then for any x there is some y such that $Iy = xy$. But $Iy = y$, so that $y = xy$, and thus y is a fixed point of x. Conversely, suppose that every element x has a fixed point y. Then $xy = y$. But $y = Iy$, so that $xy = Iy$, and thus I agrees with x on y.

    (2) Suppose that every pair of elements is compatible. Now consider any element A. Then A is compatible with I. Thus there are elements x and y such that $Ax = y$ and $Iy = x$. Since $Iy = x$, then $y = x$ (since $y = Iy$). Thus $Ax = y$ and $y = x$ and it follows that $Ax = x$, which means that x is a fixed point of A. Thus every element has a fixed point, and consequently I is agreeable by (1).

    (3) Since $Ix = x$ for every x, it follows that $II = I$, which means that I is egocentric. If I were narcissistic, that would mean that for every x, $Ix = I$. But also $Ix = x$. Hence if I were narcissistic, then we would have $x = I$ for every element x, contrary to the fact that $\mathcal{C}$ has more than one element.

30. Take M to be LI. Well, $LIx = I(xx) = xx$. Thus LI is a mocker.

31. We will see that any element θ that combines M with L must be a sage.

    We already know that $Lx(Lx)$ is a fixed point of x (by the solution to Problem 23). Now, $Lx(Lx) = M(Lx)$, and consequently $M(Lx)$ is a fixed point of x. Now, suppose that θ combines M with L. Then $\theta x = M(Lx)$, and so θx is a fixed point of x. Thus θ is a sage.

Chapter 9

Combinatorics Galore

We now discuss some of the most prominent combinators of the literature, as well as some which I introduced in *To Mock a Mockingbird*.

I. The *B*-Combinators

Introducing the B Combinator, the Bluebird

The *B*-combinator is a combinator satisfying the following condition (for all x, y and z):

$$Bxyz = x(yz).$$

This combinator is also known as the *bluebird*, a name I introduced in *To Mock a Mockingbird*. I shall sometimes use that name. This combinator is one of the basic combinators.

Problem 1. Suppose $\mathcal{C}$ contains a bluebird B and a mocker M. Then the composition condition must hold (why?). Hence, by Theorem 1 of the last chapter, every element x has a fixed point. Moreover, one can write down an expression for a fixed point of x, using just the symbols B, M and x (and parentheses, of course). Write down such an expression.

Problem 2. Show that if B and M are present (as members of $\mathcal{C}$), then some element of $\mathcal{C}$ is egocentric; in fact, write down the name of an egocentric element in terms of B and M.

Problem 3. Now write down a narcissistic element in terms of B, M and K (kestrel).

Some Derivatives of B

From B alone, one can derive combinators $D, B_1, E, B_2, D_1 B^+, D_2, \hat{E}$, which are defined by the following conditions:

(a) $Dxyzw = xy(zw)$, the dove

(b) $B_1 xyzw = x(yzw)$

(c) $Exyzwv = xy(zwv)$, the eagle

(d) $B_2 xyzwv = x(yzwv)$

(e) $D_1 xyzwv = xyz(wv)$

(f) $B^+ xyzw = x(y(zw))$

(g) $D_2 xyzwv = x(yz)(wv)$

(h) $\hat{E} xy_1 y_2 y_3 z_1 z_2 z_3 = x(y_1 y_2 y_3)(z_1 z_2 z_3)$.

All these combinators, including B, are called *compositors*, since they serve to introduce parentheses. The standard ones are the bluebird B and D, which I call the dove. We shall also have good use for E, which I call the eagle.

Problem 4. (1) Derive these combinators from B. [**Hints**: Do the following in order:

(a) Express D in terms of B.

(b) Express B_1 in terms of B and D (which can then be reduced to an expression in terms of B alone).

(c) Express E in terms of B and B_1.

(d) Express B_2 in terms of B and E.

(e) Express D_1 in terms of B and B_1, or, alternatively, in terms of B and D.

(f) Express B^+ in terms of B and D_1.

(g) Interestingly, D_2 can be expressed in terms of D alone!

(h) Express $\hat{E}$ in terms of E alone!]

(2) Using mathematical induction, show that for each positive integer n, there is a combinator B_n, derivable from B, satisfying the condition

$$B_n x y_1 \ldots y_{n+2} = x(y_1 \ldots y_{n+2}), \quad n \geq 0.$$

Note that by this definition, the bluebird B also has the name B_0, and B_1 and B_2 are the same as defined above (where different variables were used in the definition).

II. The Permuting Combinators

We now come to an interesting family of combinators known as permuters.

Introducing the Combinator T, the Thrush

The simplest permuting combinator is T, which is defined by the following:
$$Txy = yx.$$
This permuter T is standard. In *To Mock a Mockingbird*, I called it the *thrush*.

Two elements x and y are said to *commute* if $xy = yx$.

Problem 5. Prove that from the thrush T and the lark L one can derive a combinator A that commutes with every element x.

Introducing the Combinator R, the Robin

In *To Mock a Mockingbird*, I introduced a combinator R, which I termed a *robin*. It was defined by the condition:
$$Rxyz = yzx.$$
Problem 6. Show that a robin R can be derived from B and T.

Introducing the Combinator C, the Cardinal

The combinator C is defined by the condition:
$$Cxyz = xzy.$$
This C is standard in the literature, and of basic importance. I called it the cardinal in *To Mock a Mockingbird*. It is derivable form B and T, as was discovered by Alonzo Church [1941]. Church's construction was quite tricky! It used eight letters, and I doubt that it can be done with fewer.

Having derived R from B and T, it is quite simple to get C.

Problem 7. How can one derive C from R?

Problem 8. The solution of Problem 7, when reduced to B and T has nine letters. It can easily be shortened to eight, thus yielding Church's expression for C. Do this reduction.

Problem 9. Can a thrush T be derived from C and the identity combinator I?

Problem 10. We have seen that C is derivable from R alone. Can one derive R from C alone?

Note: Taking BBT for R, and RRR for C, a useful fact to observe is that for any x:
$$Cx = B(Tx)R$$
because $Cx = RRRx = RxR = BBTxR = B(Tx)R$.

Introducing the Combinator F, the Finch

A finch F is defined by the following condition:
$$Fxyz = zyx.$$
A finch F can be derived from B and T in several ways. It can be derived from B and R, or from B and C, or from T and the eagle E.

Problem 11. (a) It is easiest to derive F from all three of B, R and C. How? [Then, of course, F is derivable from B and R, or from B and C, since R and C and inter-derivable.] (b) Show how to derive F from B and E.

Introducing the Combinator V, the Vireo

A *vireo* is a combinator V satisfying the following condition:
$$Vxyz = zxy.$$
The combinator V has a sort of opposite effect to that of the finch F. V is derivable from B and T. It is easiest to derive it from C and F.

Problem 12. Derive V from C and F.

Problem 13. We have seen that V is derivable from C and F. It is also true that F is derivable from C and V. How?

Problem 14. As a curiosity, show that the identity combinator I is derivable from R and K (kestrel).

Some Relations

Given a permuting combinator A satisfying the condition $Axyz = abc$, where abc is some permutation of xyz, by A^* is meant the combinator satisfying the condition:

$$A^*wxyz = wabc.$$

Thus C^*, R^*, F^*, V^* are combinators defined by the following conditions:

$$C^*wxyz = wxzy,$$

$$R^*wxyz = wyzx,$$

$$F^*wxyz = wzyx,$$

$$V^*wxyz = wzxy.$$

Each of these combinators is derivable from B and T. It is easiest to derive them from combinators already derived from B and T.

Problem 15. Derive:
(a) C^* from B and C;
(b) R^* from B and C;
(c) F^* from B, C^* and R^*;
(d) V^* from C^* and F^*.

Consider again a permutation combinator A satisfying the condition $Axyz = abc$, where abc is a permutation of xyz. We let A^{**} be the combinator defined by the condition

$$A^{**}vwxyz = vwabc.$$

Thus C^{**}, R^{**}, F^{**}, V^{**} are combinators defined by the following conditions:

$$C^{**}vwxyz = vwxzy,$$

$$R^{**}vwxyz = vwyzx,$$

$$F^{**}vwxyz = vwzyx,$$

$$V^{**}vwxyz = vwzxy.$$

Problem 16. Show that $C^{**}, R^{**}, F^{**}, V^{**}$ are all derivable from B and T.

The problem is far simpler than it might appear. One can handle all four cases in one fell swoop!

Vireos Revisited

We have derived V from C and F, obtaining an expression, which, when reduced to B and T, has 16 letters. It can alternatively be derived from C^* and T, yielding an expression, which, when reduced to B and T, has only ten letters. How?

Problem 17. Derive V from C^* and T.

Problem 18. We have not yet considered a combinator A satisfying the condition $Axyz = yxz$. Is there one derivable from B and T?

III.   The Q-Family and the Goldfinch, G

We now come to a family of combinators, all derivable from B and T, that both permute and introduce parentheses. The first member is Q defined by the condition:

$$Qxyz = y(xz).$$

I termed Q the *queer bird* in *To Mock a Mockingbird*.

Problem 19. Derive Q from B and T. [**Hint:** There is a two-letter expression involving B and one other combinator already derived from B and T that works!]

Q_1 *and* Q_2

The two combinators Q_1 and Q_2 are defined by the following conditions:

$$Q_1xyz = x(zy),$$
$$Q_2xyz = y(zx).$$

In *To Mock a Mockingbird*, the combinators Q_1 and Q_2 were respectively called the *quixotic* and the *quizzied* birds.

Problem 20. Derive Q_1 and Q_2 from B and T, or from any combinators already derived from them.

There is an old Chinese proverb that says that if a cardinal C is present, then you cannot have a quixotic bird without a quizzied bird, nor a quizzied bird without a quixotic bird, or if there isn't such a Chinese proverb, then there should be!

Problem 21. What is the sense behind such a proverb?

The Birds Q_3 (Quirky), Q_4 (Quacky), Q_5 (Quintessential), Q_6 (Quivering)

The *quirky bird*, Q_3, the *quacky bird*, Q_4, the *quintessential bird*, Q_5, and the *quivering bird*, Q_6, are defined by the following conditions:

$$Q_3xyz = z(xy),$$
$$Q_4xyz = z(yx),$$
$$Q_5xyzw = z(xyw),$$
$$Q_6xyzw = w(xyz).$$

Problem 22. Derive these from combinators derivable from B and T.

There is, or should be, another Chinese proverb which says that if a cardinal is present, then you cannot have a quirky bird without a quacky bird, nor a quacky bird without a quirky bird.

Problem 23. What is the sense behind *that* one?

Problem 24. Show that Q_4 can be derived from T and Q_1.

Problem 25. We have seen that Q is derivable from B and T. But also, B is derivable from Q and T. Prove this. [It is not obvious.]

Problem 26. One can derive C from Q and T more easily than from B and T, in fact, with an expression of only four letters. How can this be done?

The Combinator G, the Goldfinch

In *To Mock a Mockingbird* I introduced a combinator G, called the *goldfinch*, which I found quite useful. It is defined by the condition:

$$Gxyzw = xw(yz).$$

I do not know if this combinator was previously known or not.

Problem 27. Derive G from B and T, or from combinators already derived from B and T.

IV. Combinators Derivable from B, T, M and I (λ-I Combinators)

Combinators derivable from B, T, M and I form an extremely important class, whose significance will be discussed in a later chapter. These combinators are known as λ–I combinators.

A useful combinator M_2 is defined by the condition

$$M_2 xy = xy(xy).$$

It is easily derivable from M and B.

Problem 28. Do so.

The Combinator W, the Warbler and the Combinator W', the Converse Warbler

A standard combinator is W, which is defined by the condition

$$Wxy = xyy.$$

W should not be confused with the lark L: $Lxy = x(yy)$.

Alonzo Church showed how to derive W from B, T, M and I. His derivation was both bizarre and ingenious. His expression for W involved 24 letters and 14 pairs of parentheses! Fairly soon afterwards, J. B. Rosser found an expression for W in terms of just B, T and M that included only ten letters.

I called W the *Warbler* in *To Mock a Mockingbird*. Before deriving it from B, T and M, it will prove convenient to first consider another combinator W', which might be called a *converse Warbler*. W' is defined by the condition

$$W'xy = yxx.$$

Problem 29. There are two interesting ways of deriving W' from B, T and M. One way is to first derive W' from M_2 and the robin R, and then reduce that expression to B and T, which should yield an expression of five letters. Another way is to first derive W' from B, T and M_2, which when reduced to B, T and M yields a different expression, one that also contains five letters. The reader should try to find both these expressions!

Problem 30. One can easily derive W from W' and the cardinal C, getting an expression which, when reduced to B, T and M, will have 13 letters. However, it can be further reduced to an expression having only ten letters in two different ways, depending on the choice for W. Do these things.

Problem 31. M is obviously derivable from W and T, and also from W and I. Also, I is derivable from W and K. Find these derivations.

Problem 32. Some useful relatives of W are W_1, W_2, W_3, defined by the following conditions:

$$W_1 xyz = xyzz,$$

$$W_2 xyzw = xyzww,$$

$$W_3 xyzwv = xyzwvv.$$

Show that these are all derivable from B, T and M, in fact from W and B.

The Combinator H, the Hummingbird

I have found good use for a combinator H (dubbed the hummingbird in *To Mock a Mockingbird*). The combinator H is defined by the condition

$$Hxyz = xyzy.$$

Problem 33. Show that H is derivable from B, C and W, and thus from B, M and T.

Problem 34. One can also derive W from H and R, or, more neatly, from C, H and R. How? [**Hint:** First derive W'.]

Larks Revisited

We recall the lark L defined by the condition $Lxy = x(yy)$. The lark can be derived from B, T and M in many ways.

Problem 35.
(a) Show that L is derivable from B, R and M or from B, C and M, and then reduce the latter expression to one in terms of B, M and T.
(b) Show that L is derivable from B and W. This fact is rather important.

(c) My favorite derivation of L is from M and the queer combinator Q. It is also the simplest. When reduced to B, M and T, we get the same expression as in (a). Try this!

The Combinator S, the Starling

One of the most important combinators in the literature of Combinatory Logic is the combinator S, which is defined by the following definition:

$$Sxyz = xz(yz).$$

I called S the *starling* in *To Mock a Mockingbird*. One reason why S is so important is that from S and K one can derive all possible combinators, in a sense I will make precise in a later chapter.

The starling is derivable from B, M and T, and more easily from B, C and W. The standard expression for S in terms of B, C and W has seven letters, but in *To Mock a Mockingbird*, I found another expression having only six letters. For this I used the goldfinch G, which we recall is defined by the condition $Gxyzw = xw(yz)$.

Problem 36. Derive S from B, W and G, and then reduce the resulting expression to B, C and W. [**Hint:** Use W_2, which we learned from the solution to Problem 32 can be expressed as $B(BW)$.]

We recall that the hummingbird H was defined by the condition $Hxyz = xyzy$.

Problem 37. Derive H from S and R.

Problem 38. Show that both W and M are derivable from S and T.

Problem 39. Express a warbler W in terms of S and C.

Note: We now see that the class of combinators derivable from B, C and W is the same as the class of combinators derivable from B, C and S, since S is derivable from B, C and W [by Problem 36], and W is derivable from S and C [by Problem 39].

The Order of a Combinator

By a combinator of order one is meant a combinator A such that Ax can be expressed in terms of x alone (a single variable and no combinators).

For example, M is of order 1, since $Mx = xx$, and the expression xx does not involve the letter M. Another example is the identity combinator I, since $Ix = x$. Another is WL, since $WLx = Lxx = x(xx)$.

By a combinator of order two is meant one whose defining condition involves just two variables (and no combinators), as is the case for W and L. In general, for any positive integer n, by a combinator of order n is meant one whose defining condition involves n variables. The permuting combinators C, R, F and V are of order three, as are B and Q.

It is not true that every combinator has an order, for TI cannot be of any order n, since $TIx_1, \ldots, x_n = x_1 I x_2, \ldots, x_n$, which cannot be reduced any further (and so we cannot get rid of the combinator I on the right side of this equation). On the other hand IT has an order, since it is equal to T, which is of order two.

A useful fact to note is that for any combinators A_1 and A_2, the combinator RA_1A_2 is equal to CA_2A_1, since $RA_1A_2x = A_2xA_1$, and also $CA_2A_1x = A_2xA_1$.

The P Group of Combinators

Problem 40. We shall have good use for some of the combinators P, P_1, P_2 and P_3 satisfying the following conditions:

$$\text{(a)} \quad Pxyz = z(xyz),$$
$$\text{(b)} \quad P_1xyz = y(xxz),$$
$$\text{(c)} \quad P_2xyz = x(yyz),$$
$$\text{(d)} \quad P_3xyz = y(xzy).$$

These can be derived from B, M and T, in fact from B, Q and W (and hence from B, C and W). How?

Problem 41. Show that M is derivable from P and I.

The Φ Group of Combinators

We will later need combinators Φ, Φ_2, Φ_3, Φ_4, defined by the following conditions:

$$\Phi xyzw = x(yw)(zw),$$
$$\Phi_2xyzw_1w_2 = x(yw_1w_2)(zw_1w_2),$$

$$\Phi_3 xyzw_1w_2w_3 = x(yw_1w_2w_3)(zw_1w_2w_3),$$
$$\Phi_4 xyzw_1w_2w_3w_4 = x(yw_1w_2w_3w_4)(zw_1w_2w_3w_4).$$

Problem 42. (a) Derive Φ from B and S. (b) Show by induction that for each n, there is a combinator Φ_n derivable from B and S satisfying the equation:

$$\Phi_n xyzw_1 \ldots w_n = x(yw_1 \ldots w_n)(zw_1 \ldots w_n).$$

Solutions to the Problems of Chapter 9

1. To begin with, let us note that if B is present, then the composition condition must hold, since for any elements x and y, an element that combines x with y is Bxy [since $(Bxy)z$ is $Bxyz$, which is $x(yz)$].

   We recall from the solution of Problem 1 of the last chapter that if C is any element that combines x with M, then CC is a fixed point of x (we stated this result as Theorem 1*). Well, BxM combines x with M, and so $BxM(BxM)$ is a fixed point of x. But $BxM(BxM)$ is also equal to $M(BxM)$, by the definition of M [as applied to the first M in the expansion of $M(BxM)$]. So $M(BxM)$ is also a fixed point of x (for every x!). We will use this result several times in this chapter and the next.

2. Since B is present, the composition condition holds. Then by the solution of Problem 2 of the last chapter, any fixed point of M is egocentric. Now, $M(BMM)$ is a fixed point of M [by the previous problem of this chapter, taking M for x]; hence $M(BMM)$ is egocentric. Let us double-check: To reduce clutter, let A be BMM. We are to show that MA is egocentric. Well, $MA = M(BMM) = BMM(BMM)$. But $BMM(BMM) = M(M(BMM))$, since B is a bluebird. Then we have $M(M(BMM)) = M(MA) = MA(MA)$. Thus $MA = MA(MA)$, and MA is egocentric.

3. By Problem 20 (b) of the last chapter, any fixed point of a kestrel K is narcissistic. Since $M(BKM)$ is a fixed point of K [by the solution to Problem 1], it must be narcissistic [as the reader can check directly, by showing that $M(BKM)x = M(BKM)$].

4. (1) (a) We take D to be BB. Let us check that this works: $BBxy = B(xy)$. Hence $BBxyz = B(xy)z$. Then

$$BBxyzw = B(xy)zw = xy(zw).$$

Thus $Dxyzw = xy(zw)$.

(b) We take B_1 to be DB, which can then be expressed in terms of B alone, since D can be so expressed. We can see this definition works as follows:

$$B_1xyz = DBxyz = Bx(yz).$$

Thus,
$$B_1xyzw = Bx(yz)w = x((yz)w) = x(yzw).$$

Since $D = BB$, then $DB = BBB$. Thus in terms of B alone $B_1 = BBB$.

(c) We take E to be BB_1, which in terms of B alone is $B(BBB)$. We can see this definition works as follows: $Exy = BB_1xy = B_1(xy)$. Thus

$$Exyzwv = B_1(xy)zwv = xy(zwv).$$

(d) We take B_2 to be EB (which, in terms of B, is $B(BBB)B$). Now,
$$B_2xyzw = EBxyzw = Bx(yzw).$$

Hence
$$B_2xyzwv = EBxyzwv = Bx(yzw)v = x(yzwv).$$

(e) We can take D_1 to be B_1B, or we can take D_1 to be BD. Actually B_1B is the same as BD because

$$B_1B = BBBB = B(BB) = BD.$$

It will be quicker to take D_1 to be B_1B. Then

$$D_1xyz = B_1Bxyz = B(xyz).$$

Hence,

$$D_1xyzwv = B_1Bxyzwv = B(xyz)wv = xyz(wv).$$

(f) We take B^+ to be D_1B. Thus,

$$B^+xyzw = D_1Bxyzw = Bxy(zw) = x(y(zw)).$$

(g) We take D_2 to be DD. Then

$$D_2xyzwv = DDxyzwv = Dx(yz)wv = x(yz)(wv).$$

(h) We take $\hat{E}$ to be EE. Then

$$\hat{E}\,xy_1y_2y_3z_1z_2z_3 = EExy_1y_2y_3z_1z_2z_3 = Ex(y_1y_2y_3)z_1z_2z_3$$
$$= x(y_1y_2y_3)(z_1z_2z_3).$$

(2) We will show by mathematical induction that we can define all the combinators

$$B_nxy_1\ldots y_{n+2} = x(y_1\ldots y_{n+2}), \quad n \geq 0,$$

in terms of the combinator B. Well, we have already noted that B_0 is B itself. Now assume that B_n can be defined in terms of the combinator B. Then we claim that $B_{n+1} = BBB_n$:

$$B_{n+1}xy_1y_2\ldots y_{n+2}y_{n+3} = BBB_nxy_1y_2\ldots y_{n+2}y_{n+3}$$

$$= B(B_nx)y_1y_2\ldots y_{n+2}y_{n+3} = (B_nx)(y_1y_2)\ldots y_{n+2}y_{n+3}$$

$$= x((y_1y_2)\ldots y_{n+2}y_{n+3}) = x(y_1y_2\ldots y_{n+2}y_{n+3}).$$

5. Actually, any fixed point A of T will commute with every element x. For suppose $TA = A$. Then for every element x, we see that $Ax = TAx = xA$. Thus A commutes with x.

6. Take R to be BBT. Then

$$Rxyz = BBTxyz = B(Tx)yz = Tx(yz) = yzx.$$

7. Actually C is derivable from R alone! Take C to be RRR. Then:

$$Cxyz = RRRxyz = RxRyz = Ryxz = xzy.$$

8. In terms of B and T, $C = BBT(BBT)(BBT)$, which has nine letters. It is equal to $B(T(BBT))(BBT)$, which is Church's expression.

9. Easily! Take T to be CI. Then $Txy = CIxy = Iyx = yx$.

10. Yes. Take R to be CC. Then $Rxyz = CCxyz = Cyxz = yzx$.

11. (a) BCR is a finch, because

$$BCRxyz = C(Rx)yz = (Rx)zy = Rxzy = zyx.$$

In terms of B and R, $F = B(RRR)R$. In terms of B and C, $F = BC(CC)$.

(b) $ETTET$ is a finch, because

$$ETTETxyz = TT(ETx)yz = Tx(Tyz) = Tyzx = zyx.$$

We might note that in terms of B and T, the expression BCR for the finch has 12 letters (since C has 8 and R has 3), while using the expression $B(RRR)R$ for the finch gives us 13 letters in terms of B and T; using the expression $BC(CC)$ for the finch, we get 25 letters in terms of B and T.

As for $F = ETTET$, as it stands, when E is replaced by $B(BBB)$, we get an expression of 11 letters, but the expression can be reduced to only 8 letters in terms of B and T, as we will now see:

$$ETT = B(BBB)TT = BBB(TT) = B(B(TT)).$$

Hence,

$$ETTET = B(B(TT))ET = B(TT)(ET) = B(TT)(B(BBB)T).$$

Thus, expressing F in terms of E and T, and then reducing that expression to one in terms of B and T, yields the shortest expression for the finch F.

12. Take V to be CF. Then $CFxyz = Fyxz = zxy$.

Note: When reduced to B and T, the expression CF has 16 letters (since C and F each have 8). This 16-letter expression cannot be reduced further. However, one can get an expression for V in terms of B and T which has only ten letters, as we will later see.

13. $CVxyz = Vyxz = zyx$. Thus CV is a finch, F.

14. For any element A, the element RAK must be an identity combinator, for $RAKx = KxA = x$. In particular, RKK and RRK are both identity combinators.

15. Here are the derivations requested for C^*, R^*, F^* and V^*:

   (a) Take C^* to be BC: $BCwxyz = C(wx)yz = wxzy$.
   (b) We have just derived C^* from B and C, and R^* can in fact be derived from C^* alone. Thus take R^* to be C^*C^*. Well, $C^*C^*wxy = C^*wyx$. Hence

   $$C^*C^*wxyz = C^*wyxz = wyzx = R^*wxyz.$$

   So, in terms of B and C, $R^* = BC(BC)$.
   (c) Take F^* to be BC^*R^*. Then,

   $$BC^*R^*wxyz = C^*(R^*w)xyz = R^*wxzy = wzyx.$$

(d) Take V^* to be C^*F^*. Well, $C^*F^*wxyz = F^*wyxz = wzxy$.

16. Here is the secret! Let A be any of the four combinators C, R, F, V. We can take A^{**} to be BA^*. Here is why:

$$Axyz = abc, \text{ for some permutation } abc \text{ of } xyz.$$
$$\text{Then } A^*wxyz = wabc \text{ and}$$
$$A^{**}vwxyz = BA^*vwxyz = A^*(vw)xyz = (vw)abc = vwabc.$$

17. Take V to be C^*T. Well,

$$C^*Txyz = Txzy = zxy.$$

Thus C^*T is a vireo.

18. Of course! The combinator T itself is such a combinator, since $Txy = yx$, so that $Txyz = yxz$.

19. The expression CB works:

$$CBxyz = Byxz = y(xz) = Qxyz.$$

We thus take Q to be CB.

20. Take Q_1 to be C^*B. Well, $C^*Bxyz = Bxzy = x(zy)$. Take Q_2 to be R^*B. Now, $R^*Bxyz = Byzx = y(zx)$.

21. The fact is that Q_2 is derivable from C and Q_1, and Q_1 is derivable from C and Q_2:

$$CQ_1xyz = Q_1yxz = y(zx) = Q_2xyz;$$
$$CQ_2xyz = Q_2yxz = x(zy) = Q_1xyz.$$

22. Q_3: Take Q_3 to be V^*B. Well, $V^*Bxyz = Bzxy = z(xy)$. Alternatively, we can take Q_3 to be $BC(CB)$ or QQC. In terms of B and T, we could simply take Q_3 to be BT.

Q_4: We could take Q_4 to be F^*B: We see that

$$F^*Bxyz = Bzyx = z(yx).$$

Alternately, we could take Q_4 to be

$$CQ_3: CQ_3xyz = Q_3yxz = z(yx).$$

Q_5: Take Q_5 to be $BQ: Q_3: BQxyzw = Q(xy)zw = z(xyw)$.
Q_6: Take Q_6 to be $B(BC)Q_5$. Then

$$Q_6xyzw = B(BC)Q_5xyzw = (BC)(Q_5x)yzw$$
$$= C(Q_5xy)zw = Q_5xywz = w(xyz).$$

23. Q_4 is derivable from C and Q_3, and Q_3 is derivable from C and Q_4:

$$CQ_3xyz = Q_3yxz = z(yx) = Q_4xyz.$$

$$CQ_4xyz = Q_4yxz = z(xy) = Q_3xyz.$$

24. We can express Q_4 as Q_1T:

$$Q_1Txyz = T(yx)z = z(yx) = Q_4xyz.$$

25. The bluebird B can be seen to be the same as $QT(QQ)$ as follows:

$$QT(QQ)xyz = QQ(Tx)yz = Tx(Qy)z = Qyxz = x(yz) = Bxyz.$$

26. The cardinal C can be expressed as $QQ(QT)$:

$$QQ(QT)xyz = QT(Qx)yz = Qx(Ty)z = Ty(xz) = xzy = Cxyz.$$

27. Take G to be BBC. Then

$$BBCxyzw = B(Cx)yzw = Cx(yz)w = xw(yz).$$

28. Obviously, take M_2 to be BM :

$$BMxy = M(xy) = xy(xy) = M_2xy.$$

29. For the first method, we take W' to be M_2R. Then

$$M_2Rxy = Rx(Rx)y = Rxyx = yxx = W'xy.$$

Reducing to B, T and M, we take BM for M_2 and BBT for R, obtaining the expression $BM(BBT)$.

As for the second way to derive W', we take W' to be $B(M_2B)T$. In this case,

$$B(M_2B)Txy = M_2B(Tx)y = B(Tx)(B(Tx))y$$

$$= Tx(B(Tx)y) = B(Tx)yx = Tx(yx) = yxx = W'xy.$$

Taking BM for M_2, this version of W' reduces to $B(BMB)T$. Thus we have two ways of expressing W' from B, T and M: $BM(BBT)$ and $B(BMB)T$.

30. CW' is a warbler, since $CW'xy = W'yx = xyy = Wxy$.

We now use the fact previously noted that for any x, $Cx = B(Tx)R$ (this was shown right after the statement of Problem 10 in this chapter), and so $CW' = B(TW')R$. Thus $B(TW')R$ is a warbler. Hence $B(TW')BBT$ is a warbler (since $R = BBT$). As seen in the last problem, we can take W' to be either $BM(BBT)$ or $B(BMB)T$, and so we can take W to be either

$$B(T(BM(BBT)))BBT \quad \text{or} \quad B(T(B(BMB)T))BBT.$$

The latter is Rosser's expression for W.

31.
$$WTx = Txx = xx = Mx.$$
$$WIx = Ixx = xx = Mx$$
$$WKx = Kxx = x = Ix.$$

32. We will take W_1 to be BW, W_2 to be BW_1, W_3 to be BW_2. Then
$$BWxyz = W(xy)z = xyzz = W_1xyz.$$
$$BW_1xyzw = W_1(xy)zw = (xy)zww = xyzww = W_2xyzw.$$
$$BW_2xyzwv = W_2(xy)zwv = (xy)zwvv = xyzwvv = W_3xyzwv.$$

Note: In general, if for all n we take W_{n+1} to be BW_n, then, for every n,
$$W_nxy_1 \ldots y_nz = xy_1 \ldots y_nzz$$
as can be shown by mathematical induction.

33. We take H to be W^*C^*, which is $BW(BC)$. Then
$$Hxy = W^*C^*xy = C^*xyy.$$

Thus,
$$C^*xyyz = xyzy = Hxyz.$$

34. First we take W' to be HR. Thus $HRxy = Rxyx = yxx = W'xy$. Then take W to be CW'. This is a warbler: $CW'xy = W'yx = xyy = Wxy$. Thus $C(HR)$ is a warbler, and since C is RRR, the warbler is also derivable from H and R alone.

35. (a) RMB is a lark, since $RMBxy = BxMy = x(My) = x(yy) = Lxy$. Also, so is CBM (as the reader can verify). To express RMB in terms of B, M and T, we take BBT for R in RMB, so that
$$RMB = BBTMB = B(TM)B.$$
Thus $B(TM)B$ is a lark.

   (b) BWB is also a lark, since
$$BWBxy = W(Bx)y = Bxyy = x(yy) = Lxy.$$

   (c) QM is a lark, since $QMxy = x(My) = x(yy) = Lxy$. Reducing this to B, M and T, we see that $QM = CBM$ [since $Q = CB$ by Problem 19]. Also, for any x and y, it is a fact that $Cxy = Ryx$, because [by Problem 7] $Cxy = RRRxy = RxRy = Ryx$. Thus $Cxy = Ryx$. In particular, $CBM = RMB$, so that $QM = RMB$, which is the expression we obtained for L in (a), and which reduced to $B(TM)B$.

36. Recall that W_2 was defined by the condition $W_2xyzw = xyzww$ in the statement of Problem 32 and that in the solution to Problem 32 we saw W_2 to be $B(BW)$, i.e. defined in terms of B and W. We can take the starling S to be W_2G, since $W_2Gxyz = Gxyzz = xz(yz) = Sxyz$. Since we saw in the solution to Problem 27 that $G = BBC$, we also see that S is $B(BW)(BBC)$ in terms of to B, C and W.

37. SR is a hummingbird, since $SRxyz = Ry(xy)z = xyzy = Hxyz$.

38. (a) ST is a warbler, since $STxy = Ty(xy) = xyy = Wxy$.
    (b) STT is a mocker, since $STTx = Tx(Tx) = Txx = xx = Mx$. Another way to see this is based on our having seen that WT in a mocker [in the solution to Problem 31], and since we can take ST for W [because $STxy = Ty(xy) = xyy = Wxy$], again we see that STT is a mocker.

39. By Problem 34, $C(HR)$ is a warbler W. By the solution to Problem 37, SR is a hummingbird (H). Thus $C(SRR)$ is a warbler. But by the solution to Problem 10, CC is a robin R, and so, for W, we can take $C(S(CC)(CC))$.

40. (a) Take P to be W_2Q_5. Then $W_2Q_5xyz = Q_5xyzz = z(xyz) = Pxyz$.
    (b) Take P_1 to be LQ. Then $LQxyz = Q(xx)yz = y(xxz) = P_1xyz$.
    (c) Take P_2 to be CP_1. Then $CP_1xyz = P_1yxz = x(yyz) = P_2xyz$.
    (d) Take P_3 to be BCP. Then

$$BCPxyz = C(Px)yz = Pxzy = y(xzy) = P_3xyz.$$

41. Take M to be PII. Then $PIIx = x(IIx) = x(Ix) = xx = Mx$.

42. (a) Take Φ to be $B(BS)B$. Then

$$B(BS)Bxyzw = BS(Bx)yzw = S(Bxy)zw$$

$$= Bxyw(zw) = x(yw)(zw) = \Phi xyzw.$$

(b) We take $\Phi_1 = \Phi$. Now, suppose Φ_n is defined and behaves as shown in the equation $\Phi_n xyzw_1 \ldots w_n = x(yw_1 \ldots w_n)(zw_1 \ldots w_n)$. We take Φ_{n+1} to be $B\Phi_n\Phi$. Then

$$B\Phi_n\Phi xyzw_1 \ldots w_n w_{n+1} = \Phi_n(\Phi x)yzw_1 \ldots w_n w_{n+1}$$

$$= (\Phi x)(yw_1 \ldots w_n)(zw_1 \ldots w_n)w_{n+1}$$

$$= x(yw_1 \ldots w_{n+1})(zw_1 \ldots w_{n+1})$$

$$= \Phi_{n+1}xyzw_1 \ldots w_{n+1}.$$

Chapter 10

Sages, Oracles and Doublets

Sages Again

We recall that by a *sage* is meant a combinator θ such that θx is a fixed point of x for every x; in other words, $x(\theta x) = \theta x$.

Several papers in the literature are devoted to the construction of sages.

In the chapter before last, we did not construct any sages. We merely proved that a sage exists, providing that the mocker M exists and that the composition condition holds. We now turn to several ways of constructing sages.

Problem 1. Construct a sage from B, M and the robin R, recalling that $Rxyz = yzx$. [**Hint**: Recall from the solution to Problem 1 in Chapter 9 that $M(BxM)$ is a fixed point of x.]

Problem 2. Now derive a sage from B, M and the cardinal C.

Problem 3. A simpler construction of a sage uses M, B and the lark L. One such construction provides simple alternative solutions to the last two problems. Can you find it?

Problem 4. Derive a sage from B, M and W.

It is a bit tougher to derive a sage from B, W and C. We will consider some ways of doing so.

Problem 5. Let us first derive a sage from B, L and the queer bird Q [$Qxyz = y(xz)$].

Problem 6. Using the result of the last problem, now write down an expression for a sage in terms of B, W and C.

Problem 7. A particularly neat and simple construction of a sage is from Q, M and L. Can you find it?

Problem 8. Actually, a sage can be constructed from Q and M alone. How?

We recall the starling S satisfying the condition $Sxyz = xz(yz)$.

Problem 9. Construct a sage from L and S.

Problem 10. Now show that a sage can be constructed from B, W and S. This can be done in several ways, one of which uses only five letters. (The sage presented in the set of solutions with 5 letters is due to Curry.)

Problem 11. Construct a sage from M and P. $[Pxyz = z(xyz)]$.

Problem 12. Construct a sage from P and W.

Problem 13. Construct a sage from P and I.

The Turing Combinator

In 1937, the great Alan Turing discovered a remarkable combinator U, which will soon surprise you! It is defined by the condition:

$$Uxy = y(xxy).$$

It is ultimately derivable from B, T and M in several ways (it is also derivable from S, L and I).

Problem 14. Derive U from W_1 and P_1.

Problem 15. Derive U from W and P.

Problem 16. Derive U from P and M.

Problem 17. Derive U from S, I and L.

Now for the surprise!

Problem 18. There is a sage that is derivable from U alone! How?

The Oracle O

A truly remarkable combinator O is defined by the condition:

$$Oxy = y(xy).$$

You will soon see why O is so remarkable! In *To Mock a Mockingbird*, I dubbed it the *owl*, but I now prefer the grander name *oracle*, which is befitting it.

Problem 19. Derive O from Q and W.

Problem 20. Derive O from B, C and W.

Problem 21. Derive O from S and I.

Problem 22. Derive O from P and I.

Problem 23. Derive a sage from O, B and M.

Problem 24. Derive a sage from O and L.

Problem 25. Derive a Turing combinator U from O, B and M.

Problem 26. Derive a Turing combinator from O and L.

Now, why is the oracle O so remarkable? Well, in the solutions to Problems 23 and 24, we have seen two sages constructed from O and other combinators, and in both cases the sages happened to be fixed points of O. Was that a mere coincidence? No! The first remarkable fact about the oracle O is that all of its fixed points are sages.

Problem 27. Prove that all fixed points of O are sages.

Now for the second remarkable fact about O. The converse of the statement of Problem 27 holds, i.e., all sages are fixed points of O. Thus the class of all fixed points of O is the same as the class of all sages!

Problem 28. Prove that every sage is a fixed point of O.

Double Sages

A pair (y_1, y_2) is called a *double fixed point* of a pair (x_1, x_2) if $x_1 y_1 y_2 = y_1$ and $x_2 y_1 y_2 = y_2$. We say the system has the *weak double fixed point property* if every pair (x_1, x_2) has a double fixed point. We say that the system had the *strong double fixed point property* if there is a pair (θ_1, θ_2), called *double sage pair* — or, more briefly, a *double sage* — such that for every x and y, the pair $(\theta_1 xy, \theta_2 xy)$ is a double fixed point of (x, y), i.e.

$$x(\theta_1 xy)(\theta_2 xy) = \theta_1 xy$$

and
$$y(\theta_1 xy)(\theta_2 xy) = \theta_2 xy.$$

There are many ways to construct double sages from the combinators B, M and T, several of which can be found in Chapter 17 of my book *Diagonalization and Self-Reference* [Smullyan, 1994], and one of which we shall now consider. Others, which I will leave as exercises, have solutions which can be found in the book just mentioned.

The Nice Combinator N

My favorite construction of a double sage uses a combinator N satisfying the following condition:

$$Nzxy = z(Nxxy)(Nyxy).$$

The existence of such an N obviously implies that the system has the *weak* double fixed point property.

Problem 29. Why?

Before considering the construction of a double sage, let us see how a nice combinator N can be ultimately derived from B, T and M. To this end, we will use a combinator n defined by the condition:

$$nwzxy = z(wxxy)(wyxy).$$

Problem 30.
(a) Derive n from B, C and W or from combinators derivable from them. [It is easiest to derive n directly from C, W, W^*, V^* and Φ_3.]
(b) Now show that any fixed point of n (such as $M(Ln)$) is a nice combinator.

Problem 31. Show that from N, C, W and W_1 one can construct combinators θ_1 and θ_2 such that (θ_1, θ_2) is a double sage.

Double Oracles

We will call a pair (A_1, A_2) a *double oracle* if every double fixed point of (A_1, A_2) is a double sage. It turns out that every double sage is a double fixed point of any double oracle (A_1, A_2).

To my delight, double oracles exist, if B, T and M are present.

Let O_1 and O_2 be defined by the following conditions:

$$O_1 zwxy = x(zxy)(wxy),$$
$$O_2 zwxy = y(zxy)(wxy).$$

Problem 32. Show that (O_1, O_2) is a double oracle.

Solutions to the Problems of Chapter 10

1. Since $BxM(BxM) = M(BxM)$ is a fixed point of x (by the solution to Problem 1 of Chapter 9), we want to find a combinator θ such that $\theta x = M(BxM)$. Well, by the definition of $R, BxM = RMBx$, and so $M(BxM) = M(RMBx)$. Also, by the definition of B,

$$M(RMBx) = BM(RMB)x.$$

We thus take θ to be $BM(RMB)$.

2. Let us now use a fact stated at the end of the last chapter, namely that for any combinators A_1 and A_2, the combinator RA_1A_2 is equal to CA_2A_1. In particular, RMB is equal to CBM, so that $RMBx = CBMx$. Then, since $M(RMBx)$ is a fixed point of x, so is $M(CBMx)$. Also, $M(CBMx) = BM(CBM)x$. We thus now take θ to be $BM(CBM)$.

3. We now use the fact that $Lx(Lx)$ is a fixed point of x. Thus so is $M(Lx)$, which is also $BMLx$. Thus BML is a sage.

We proved in the last chapter (in the solution to Problem 35) that RMB is a lark, so that CBM is also a lark, since it is equal to RMB. We can replace the L in BML with either RMB or CBM, showing that both $BM(RMB)$ and $BM(CBM)$ are sages.

This shows us alternative ways of solving Problems 1 and 2.

4. As we saw in the solution of Problem 35 of the last chapter, BWB is also a lark. Thus we can also replace the L In BML by BWB, showing us that $BM(BWB)$ is also a sage.

5. We again use the fact that $Lx(Lx)$ is a fixed point of x. Now,

$$W(QL(QL))x = QL(QL)xx = QL(Lx)x = Lx(Lx).$$

Thus $W(QL(QL))$ is a sage.

6. We have just proved that $W(QL(QL))$ is a sage. We can take CB for Q (by the solution to Problem 19 in Chapter 9), thus getting the expression $W(CBL(CBL))$, which can be shortened to $W(B(CBL)L)$. Then, by the solution of Problem 35 of Chapter 9, we can take BWB for L, thus getting the sage

$$W(B(CB(BWB))(BWB)).$$

We will later find another solution to this problem.

7. Again we use the fact that $Lx(Lx)$, and hence $M(Lx)$, is a fixed point of x. But $M(Lx) = QLMx$. Thus QLM is a sage.

8. In the expression QLM of the previous problem, we can take QM for L (by Problem 35 of Chapter 9), thus obtaining $Q(QM)M$ as another sage.

9. $Lx(Lx)$ is a fixed point of x. But $SLLx = Lx(Lx)$. Hence SLL is a sage!

10. Here is one way to get a sage from S, W and B: Since SLL is a sage (as shown in the solution to the last problem), we can take BWB for L in SLL, obtaining $S(BWB(BWB))$, which can be shortened to $S(W(B(BWB)))$, which has six letters. However, there is a more clever and more economical way: By the definition of W, $SLL = WSL$. Now replace L by BWB in WSL, and we have the sage $WS(BWB)$, one letter shorter! This is Curry's expression for a sage.

Note that since S is derivable from B, C and W (by Problem 36 of Chapter 9), so is $WS(BWB)$, thus getting a third way of deriving a sage from B, C and W.

11. $M(PM)$ is a sage: $M(PM)x = PM(PM)x = x(M(PM)x)$.

12. $WP(WP)$ is a sage: $WP(WP)x = P(WP)(WP)x = x((WP)(WP)x)$.

13. $PII(P(PII))$ is a sage, since $M(PM)$ is a sage by the solution to Problem 11 above, and M can be expressed by PII, as seen in the solution to Problem 41 of Chapter 9.

14. W_1P_1 is a Turing combinator: $W_1P_1xy = P_1xyy = y(xxy) = Uxy$.

15. Another Turing combinator is WP, since
$$WPxy = Pxxy = y(xxy) = Uxy.$$

16. Another Turing combinator is PM, since
$$PMxy = y(Mxy) = y(xxy) = Uxy.$$

17. Another Turing combinator is $L(SI)$, since
$$L(SI)xy = SI(xx)y = Iy(xxy) = y(xxy) = Uxy.$$

18. In the equation $Uxy = y(xxy)$, just substitute U for x, and we have $UUy = y(UUy)$. Thus UU is a sage.

19. QQW is an oracle, since $QQWxy = W(Qx)y = Qxyy = y(xy) = Oxy$.

20. We take CB for Q in QQW from the previous problem, which shows us that $CB(CB)W$ is an oracle.

21. SI is an oracle, since $SIxy = Iy(xy) = y(xy) = Oxy$.

22. PI is an oracle, since $PIxy = y(Ixy) = y(xy) = Oxy$.

23. See the solution to Problem 25 below.

24. See the solution to Problem 26 below.

25. BOM is a Turing combinator, since
$$BOMxy = O(Mx)y = y(Mxy) = y(xxy) = Uxy.$$

Now, since BOM is a Turing combinator, and if U is any Turning combinator, then UU is a sage (by the solution to Problem 18), it follows that $BOM(BOM)$ is a sage (which can also be verified directly), which solves Problem 23.

Please note that $BOM(BOM)$ is also a fixed point of O. Indeed (as noted in the solution to Problem 1 of Chapter 9 and recalled in the solution to Problem 1 of this chapter), $BxM(BxM)$ is a fixed point of x, for any x.

26. LO is a Turing combinator, since $LOxy = O(xx)y = y(xxy) = Uxy$. Now, since LO is a Turing combinator, then $LO(LO)$ is a sage, which solves Problem 24. [Now we can say, "Lo! Lo! A sage!!"] Again $LO(LO)$ is not only a sage, but also a fixed point of O!

27. Suppose A is a fixed point of O. Then $OA = A$. Hence

$$Ax = OAx = x(Ax).$$

Since $Ax = x(Ax)$, A is a sage.

28. Consider any sage θ. For any x, we have $\theta x = x(\theta x)$. Also $O\theta x = x(\theta x)$. Thus $\theta x = O(\theta x)$ (they are both equal to $x(\theta x)$). Since $\theta x = O\theta x$ for every term x, it follows that $\theta = O\theta$, which means that θ is a fixed point of O!

29. In the equation $Nzxy = z(Nxxy)(Nyxy)$ substitute x for z to obtain:
    (1) $Nxxy = x(Nxxy)(Nyxy)$.

    Next, in the equation $Nzxy = z(Nxxy)(Nyxy)$ substitute y for z to obtain:
    (2) $Nyxy = y(Nxxy)(Nyxy)$.

    By (1) and (2), the pair $(Nxxy, Nyxy)$ is a double fixed point of (x, y).

30. (a) We take n to be $C(V^*\Phi_3 W(W_2 C))$. We leave it to the reader to verify that
    $$nwzxy = z(wxxy)(wyxy).$$

    (b) Let N be any fixed point of n. Then $N = nN$. Thus
    $$Nzxy = nNzxy = z(Nxxy)(Nyxy).$$

31. Take θ_1 to be WN and θ_2 to be $W_1(CN)$ Then
    (a) $\theta_1 xy = WNxy = Nxxy$, so that $\theta_1 xy = Nxxy$.
    (b) $\theta_2 xy = W_1(CN)xy = CNxyy = Nyxy$, so that $\theta_2 xy = Nyxy$.

    Now,
    (1) $Nxxy = x(Nxxy)N(yxy)$ and
    (2) $Nyxy = y(Nxxy)N(yxy)$.

    Since $Nxxy = \theta_1 xy$ and $Nyxy = \theta_2 xy$, then, by (1) and (2), we have:
    (1') $\theta_1 xy = x(\theta_1 xy)(\theta_2 xy)$ and
    (2') $\theta_2 xy = y(\theta_1 xy)(\theta_2 xy)$.

    Thus (θ_1, θ_2) is a double sage.

32. Suppose (A_1, A_2) is a double fixed point of (O_1, O_2). Then
    $$O_1 A_1 A_2 = A_1$$
    and $O_2 A_1 A_2 = A_2$. Hence:
    $$A_1 xy = O_1 A_1 A_2 xy = x(A_1 xy)(A_2 xy),$$
    $$A_2 xy = O_2 A_1 A_2 xy = y(A_1 xy)(A_2 xy).$$

Thus (A_1, A_2) is a double sage, and (O_1, O_2) is a double oracle, as we were asked to show.

Before proving that every double sage is a double fixed point of (O_1, O_2), let us note that if $A_1 xy = A_2 xy$ for all x and y, then $A_1 = A_2$, because from $A_1 xy = A_2 xy$ for all x and y it follows that $A_1 x = A_2 x$ for all x, and hence that $A_1 = A_2$.

Now, suppose that (θ_1, θ_2) is a double sage. Thus:

(1) $\theta_1 xy = x(\theta_1 xy)(\theta_2 xy)$,

(2) $\theta_2 xy = y(\theta_1 xy)(\theta_2 xy)$.

Also,

(1′) $O_1 \theta_1 \theta_2 xy = x(\theta_1 xy)(\theta_2 xy)$,

(2′) $O_2 \theta_1 \theta_2 xy = y(\theta_1 xy)(\theta_2 xy)$.

Therefore, $\theta_1 xy = O_1 \theta_1 \theta_2 xy$ and $\theta_2 xy = O_2 \theta_1 \theta_2 xy$.

Then $\theta_1 = O_1 \theta_1 \theta_2$ and $\theta_2 = O_2 \theta_1 \theta_2$, which means that (θ_1, θ_2) is a double fixed point of (O_1, O_2).

Chapter 11

Complete and Partial Systems

I. The Complete System

From the two combinators S and K, all possible combinators are derivable! Let me explain just what I mean by this.

In the formal language of combinatory logic, we have a denumerable list $x_1, x_2, \ldots, x_n, \ldots$ of symbols called *variables*, and some symbols called *constants*, each of which is the name of an element of the applicative system. The notion of *term* is recursively defined by the following conditions:

(1) Every variable and constant is a term.

(2) If t_1 and t_2 are terms, so is $(t_1 t_2)$.

It is understood that no expression is a term unless its being so is a consequence of conditions (1) and (2) above.

Problem 1. Give a more explicit definition of what it means for an expression to be a term.

We continue to abbreviate terms by removing parentheses if no ambiguity can result, using the afore-mentioned convention that parentheses are to be restored to the left, e.g. xyz is to be read $((xy)z)$ etc.

For a moment, let us recall in general all the combinators you have been introduced to in Chapters 8 through 10. You learned each combinator's meaning from what we have called its "defining equation" (its defining condition). Most of those defining equations for combinators are of the form $Ax_1 x_2 \ldots x_n = t(x_1 x_2, \ldots, x_n)$, where $x_1 x_2, \ldots, x_n$ are the variables that occur in the term t and the term t is built only from these variables

215

(and parentheses; we will now [mostly] stop mentioning those necessary parentheses, which are always there whenever a term is not a single variable or a single constant, even if we suppress writing some of them for easier reading). For instance, you were introduced to the robin by its defining equation $Rxyz = yzx = ((yz)x)$ and to the cardinal by its defining equation: $Cxyz = xzy = ((xz)y)$. However, the defining equation of the kestrel, $Kxy = x$ is different in that not all the variables that we wish to have as arguments of the combinator K occur in the term t (which is x in this case), although all the variables in t occur in the list of arguments for K, the combinator being defined.

Combinators A of the first type, in which the variables of the term and the combinator being defined are identical, are called non-cancellative combinators (or λ-I combinators). We will discuss this subgroup of combinators further in Section II. Combinators A of the second type, in which not all the variables in the arguments of the combinator being defined occur in the term t are called *cancellative combinators*, because A effectively cancels out those of the variables that do not occur in t. For example, the bluebird has the defining equation $Bxyz = x(yz)$ in which the variables are identical on both sides of the equation, so it is non-cancellative. But in $Kxy = x$ and $(KI)xy = y$, one of the two arguments on the left is missing in the term on the right, so the two combinators K and KI are cancellative (these are the only cancellative operators you have seen so far).

Thus the general form of a defining equation for a combinator is $Ax_1x_2\ldots x_n = t$, where t is constructed only from variables and the variables of t are *among* $x_1, x_2, \ldots, x_n$. I.e. although t must contain at least one variable in order to be a term, and every variable in t is one of the variables $x_1, x_2, \ldots, x_n$, not all the variables that occur in $x_1, x_2, \ldots, x_n$ need to occur in t (e.g. as in the case of the kestrel).

When t is a term constructed only from variables that are among $x_1, x_2, \ldots, x_n$, and our applicative system $\mathcal{C}$ contains an element A (no element of $\mathcal{C}$ contains variables) such that $Ax_1x_2\ldots x_n = t$ holds, then we say that A is a *combinator* of the applicative system $\mathcal{C}$ and also say that the combinator A with arguments $x_1, x_2, \ldots, x_n$ *realizes* the term t. The equation $Ax_1x_2\ldots x_n = t$ is called the *defining equation* of the combinator A.

It turns out that all *constants* in any applicative system are *combinators*, because in applicative systems the meaning of constants is always specified by a defining equation. And the systems are also always defined so that every term constructed from constants alone is assumed to be a combinator of the system. Thus the combinators and elements of the system are one and the same. But the constants may be a subset of the set of elements.

So now you know what I mean by "every possible combinator" when I said that every possible combinator is derivable from S and K. I was talking about any combinator that can be specified by a defining equation of the above form!

Thus the claim that every combinator in our applicative system C can be derived from (constants) S and K just says that for every combinator A specified by a defining equation $Ax_1x_2 \ldots x_n = t$ (where t is built only from variables, and those variables are among $x_1, x_2, \ldots, x_n$), we can find a term $A_{S,K}$ constructed only from S and K (and no variables at all!) such that we can replace the A in its defining equation by $A_{S,K}$ and get a true equation (true because of the meanings of S and K): $A_{S,K}x_1x_2 \ldots x_n = t$.

Consequently, if we assume our applicative system C includes the elements (combinators) S and K, and we can prove our claim, then *every* defining equation you have seen that has defined one of the combinators you were introduced to, specifies a combinator that can be derived only from S and K, so that the only constants we need in our system to obtain all the combinators you have seen (and infinitely many more) are S and K.

But why should a specification of the ordered arguments of a combinator and a term built only from variables included in those arguments be enough to specify *any* combinator? Well, first of all, as you have just seen, such a specification includes an infinite number of combinators, including all you have studied, as well as all those that can be defined from *any* term built from variables, and an ordered list of argument variables that includes the variables of the term. The claim that these are "all the combinators" essentially says that the essence of combinators is hidden in those terms built from variables and the argument list for the combinator. Here's one way to think about that: Since applicative systems are to be imagined as being concerned with the order of "applications" of some sort of things to other things of the same sort (say, applications involving functions, algorithms, processes, etc.; you will see some quite varied interpretations of applications in the next chapter), *the structure of parentheses in complex terms involving variables can be seen as defining an order on the applications* of whatever is filled in for the variables at some point in time. For instance, consider the order of applications defined by the parentheses of $(((rs)((tu)v))((wx)y))$. [It certainly has nothing to do with the alphabetical order of the names of the variables.] In any case, you can see that, if we can prove what we have claimed, then in any applicative system including at least S and K as constants, there is a combinator A for which

$$Arstuvwxy = (((rs)((tu)v))((wx)y)),$$

as well as a (cancellative) combinator A' for which

$$A'rstuvwxy = ((wu)r)(((vx)u)s).$$

Combinatorial Completeness

Now we wish to prove our principal claim, namely that given any ordered list of variables to be taken as arguments for a combinator, and any term t built only from variables included in the list of arguments, there is a combinator with the given arguments (in the order specified in the list) that is derivable from just S and K (no variables) which realizes the term t.

To begin with, the identity combinator I is derivable from S and K: just take I to be SKK, for $SKKx = Kx(Kx) = x = Ix$. Thus it suffices to show that every combinator is derivable from S, K and I. This is what we will now do.

Let us first mention the following induction principle, which we will call the *term induction principle*: Consider a set S whose members are variables or constants or both, and let S^+ be the set of all terms built from elements of S (and parentheses). To show that all terms in S^+ have a certain property P, it suffices to show that all elements of S have the property, and that for any terms t_1 and t_2 of S^+, if t_1 and t_2 have property P, then so does (t_1t_2). This principle is as self-evident as mathematical induction, and can in fact be derived from mathematical induction on the number n of occurrences of elements of S in the term.

Let us call $\mathcal{W}$ a *simple term* if it is built only from variables and S, K, and I. In the algorithm we are about to present, *we will assume we will be starting with a simple term W built only from variables*, but we will successively turn it into a simple term built from S, K, and I and variables. In the end, we will have built a term $\mathcal{T}$ from $\mathcal{W}$ that is built only of S, K, and I (by having eliminated all the variables in a specified ordered list of arguments for a combinator $\mathcal{T}$, a list that includes all the variables in $\mathcal{W}$). And if $x_1, \ldots, x_n$ is the list of variables specified as arguments for the combinator, it will be the case that $\mathcal{T}x_1 \ldots x_n = \mathcal{W}$, where the $\mathcal{W}$

in question is the original one containing only variables (all of which are included among the $x_1, \ldots, x_n$).

Now, consider any simple term W and consider a variable x, which may or may not be in W. We shall call a term W_x an x-eliminate of W if the following three conditions hold:

(a) The variable x does not occur in W_x.

(b) All variables in W_x are in W.

(c) The equation $W_x x = W$ holds.

We wish to first show that every simple term W has an x-eliminate W_x that is derivable from S, K and I and variables in W excluding the variable x. We define W_x by the following inductive rules:

(1) If W is x itself, we take W_x to be I.

(2) If W is any term in which x does not occur, we take W_x to be KW.

(3) If W is a compound term WV in which x occurs, we take $(WV)_x$ to be $SW_x V_x$.

Note that for any possible simple term, one and only one of the three rules will apply. When rule (3) is applied, it will require finding the x-eliminates of the two parts of the compound term.

When we called the scheme of rules for x-elimination we just introduced inductive, we meant that if we have a simple term from which we wish to eliminate the variable x we operate repeatedly on the components of our term working from outside inwards looking for x-eliminates of the components of the term, until we have eliminated x from all the subterms and finally the whole term.

Let us consider an example. Let's see if we can derive from S, K and I a combinator A of one argument x with the defining equation $Ax = (xx)$. (Thus we are checking whether the mocker/mockingbird is derivable from S, K and I.) We want to see if eliminating x, the only variable in the term (and the only variable in the list of required arguments for the combinator) will give us a term in S, K and I. Well, (xx) is a compound term in which x occurs, so rule (3) tell us that the x-eliminate of (xx) consists of the starling S placed in front of the x-eliminates of the two parts of the compound term, which are both x. By rule (1), the x-eliminate of x is I. So the x-eliminate of (xx) is SII. Let us check that the three conditions which an x-eliminate must satisfy actually hold here: (a) The variable x does not occur in SII; (b) All variables in SII are in (xx) [this holds vacuously, since there are no variables in SII]; (c) $SIIx = Ix(Ix) = xx$.

(Of course xx is just (xx) with its outer parentheses suppressed for easier reading.)

Now let us prove that for any simple term, the method of x-eliminates we have described does result in a term that satisfies the conditions on an x-eliminate.

Proposition 1. $\mathcal{W}_x$ *is an x-eliminate of $\mathcal{W}$. In other words*:
(1) *The variable x does not occur in $\mathcal{W}_x$.*
(2) *All variables of $\mathcal{W}_x$ occur in $\mathcal{W}$.*
(3) *The equation $\mathcal{W}_x x = \mathcal{W}$ holds.*

This proposition is proved by term induction.

Problem 2. (i) Show that if $\mathcal{W}$ is x, then $\mathcal{W}_x$ is an x-eliminate of $\mathcal{W}$.

(ii) Show that if $\mathcal{W}$ is any term in which x does not occur, then $\mathcal{W}_x$ is an x-eliminate of $\mathcal{W}$.

(iii) Show that if $\mathcal{W}_x$ and $\mathcal{V}_x$ are x-eliminates of $\mathcal{W}$ and $\mathcal{V}$ respectively, so is $(\mathcal{W}\mathcal{V})_x$ (i.e. so is $S\mathcal{W}_x\mathcal{V}_x$).

It then follows by term induction that the term $\mathcal{W}_x$ is an x-eliminate of $\mathcal{W}$ for every simple term $\mathcal{W}$.

Problem 3. Since I is an x-eliminate of x, it follows that an x-eliminate of $\mathcal{W}x$ is $S\mathcal{W}_x I$. However, if x does not occur in $\mathcal{W}$, there is a much simpler x-eliminate of $\mathcal{W}x$, namely $\mathcal{W}$ itself! Prove this.

Now the reader might already realize that if the term t from which we wish to "eliminate" variables contains a number of variables, we must successively "eliminate" each of the variables that occur in the term, if we wish to arrive at a variable-free term (i.e. the desired combinator). Indeed, we must also "eliminate" all the variables that do not occur in our original term, but which occur in the ordered list of argument variables for the combinator we are seeking. Moreover, we must do our variable eliminations in a particular order. To that process we now turn.

For two variables x and y, by $\mathcal{W}_{x,y}$ is meant $(\mathcal{W}_x)_y$.

We shall call a term $\mathcal{T}$ an x, y eliminate of $\mathcal{W}$ if the following three conditions hold:
(1) Neither of the variables x, y occurs in $\mathcal{T}$.
(2) All variables of $\mathcal{T}$ occur in $\mathcal{W}$.
(3) The equation $\mathcal{T}xy = \mathcal{W}$ holds.

Problem 4. Which, if either, of the following statements is true?

(1) $\mathcal{W}_{x,y}$ is an x, y eliminate of $\mathcal{W}$.

(2) $\mathcal{W}_{x,y}$ is an y, x eliminate of $\mathcal{W}$.

We now take $\mathcal{W}_{x_1 x_2 \ldots x_n}$ as an abbreviation of $(\ldots (\mathcal{W}_{x_1})_{x_2}) \ldots)_{x_n})$.

We define a term $\mathcal{T}$ to be an $x_1, x_2, \ldots, x_n$ eliminate of $\mathcal{W}$ if the following three conditions hold:

(1) None of the variables $x_1, x_2, \ldots, x_n$ occur in $\mathcal{T}$.

(2) All the variables in $\mathcal{T}$ occur in $\mathcal{W}$.

(3) The equation $\mathcal{T} x_1 x_2 \ldots x_n = \mathcal{W}$ holds.

From the result of Problem 4, and mathematical induction, we have:

Proposition 2. $\mathcal{W}_{x_n x_{n-1} \ldots x_1}$ *is an* $x_1, x_2, \ldots, x_n$ *eliminate of* $\mathcal{W}$.

Now let $\mathcal{W}$ be a term built only from variables among $x_1, x_2, \ldots, x_n$. Suppose we want a combinator A with arguments $x_1 x_2 \ldots x_n$ derivable from only S, K and I (using no variables) and satisfying the condition $A x_1 \ldots x_n = \mathcal{W}$. Well, we take A to be $\mathcal{W}_{x_n x_{n-1} \ldots x_1}$, which is an $x_1, x_2, \ldots, x_n$ eliminate of $\mathcal{W}$.

All variables of A occur in $\mathcal{W}$, hence are among $x_1, x_2, \ldots, x_n$, yet none of the variables $x_1, x_2, \ldots, x_n$ are in A, which means that A contains no variables at all. Since A has been formed by a variable-elimination scheme which introduces only the combinators S, K, and I, A must be built only from S, K, and I. And the condition $A x_1 x_2 \ldots x_n = \mathcal{W}$ holds.

Now we define a class of combinators to be *combinatorially complete* when, for every simple term t and every list of possible arguments $x_1, x_2, \ldots, x_n$ for a combinator, if every variable in the term t occurs in $x_1, x_2, \ldots, x_n$, there is a combinator A in the class such that $A x_1 x_2 \ldots x_n = t$ (i.e. when, for any such term t, there is a combinator A with arguments $x_1, x_2, \ldots, x_n$ that realizes t).

Thus we have proved:

Theorem. *The class of combinators derivable from* S *and* K *is combinatorially complete.*

We would like to point out that, given a term t in S, K and I (that is, including terms in which variables are liberally strewn throughout) and an ordered list of variables $x_1, x_2, \ldots, x_n$ including all the the variables in t, the method of eliminates described here can also be used, if one wishes, to transform any such term t into an expression $\mathcal{W}x_1x_2\ldots x_n$ in which $\mathcal{W}$ is a combinator built only with S, K and I. (To see this one only has to look once more through the method and the proof that the method works to see that it applies as well in this situation.)

Now let us consider an example of how to find a cancellative combinator with more than one variable using the method of eliminates: Suppose we want a combinator A satisfying the condition $Axy = y$. Thus A has the two arguments x, y and the term we want to use to define A does not include x. Well, we want A to be a y, x eliminate of y, i.e. an x-eliminate of a y-eliminate of y. Thus we must first find a y-eliminate of y, and that is simply I. We then want an x-eliminate of I, which is KI. Thus our solution should be KI. Let us check: $KIx = I$, so that $KIxy = Iy = y$.

Now for a permuting combinator: How can we derive the thrush T from S, K and I? $[Txy = yx]$. We have to find an x-eliminate of a y-eliminate of yx. Well, a y-eliminate of y is I, and a y-eliminate of x is Kx, and so a y-eliminate of yx is $SI(Kx)$. We now need an x-eliminate of $SI(Kx) = (SI)(Kx)$. Well, an x-eliminate of SI is $K(SI)$, and an x-eliminate of Kx is simply K (by Problem 3), and so an x-eliminate of $SI(Kx)$ is $S(K(SI))K$. Thus $S(K(SI))K$ should be our solution. Let us check:

$$S(K(SI))Kx = K(SI)x(Kx) = SI(Kx).$$

Then $S(K(SI))Kxy = SI(Kx)y = Iy(Kxy) = y(Kxy) = yx$.

It would be a good exercise for the reader to try to derive various combinators for M, W, L, B, C from S, K and I, using the procedure of eliminates. The procedure can be easily programed for a computer. However, it should be pointed out that the procedure, surefire as it is, can be very tedious and often leads to much longer expressions than can be found by using some cleverness and ingenuity.

There is an interesting table on Internet called "Combinator Birds" that gives an expression in S and K alone (no I) for all the combinators you have encountered in this book and more. This table can currently be found at angelfire.com/tx4/cus/combinator/birds.html. For instance, there you will see how exceedingly long are the derivations from S and K alone of the finch and the hummingbird. In the first column of the table it is immediately after the Greek letter λ (and before the period) that the creators of the table list the arguments of the combinator in question, since those arguments cannot be determined simply from the constant-free term used to define the combinator. For the goldfinch, the first column entry in the goldfinch row, $\lambda abcd \cdot ad(bc)$, tells us that the combinator referred to in this row of the table has the arguments a, b, c and d, and the defining equation is $Gabcd = ad(bc)$. The next two column entries for the row tells us that the common symbol for this combinator is G and that it is commonly called the goldfinch. The next column entry on the row tells us that this combinator can be derived from B, C and W as BBC. And then the final column entry for the goldfinch row tells us the goldfinch can be derived from S and K alone by the term

$$((S(K((S(KS))K)))((S((S(K((S(KS))K)))S))(KK)).$$

A Fixed Point Problem

There are two results of combinatory logic known as the *First* and *Second Fixed Point Theorems*. The second one will be dealt with in the next chapter. The first will be considered now. To motivate this, the reader should try the following exercise:

Exercise. (a) Find a combinator Γ that satisfies the condition $\Gamma xy = x(\Gamma y)$. (b) Find a combinator Γ that satisfies the condition $\Gamma xy = \Gamma yx$.

The solutions of (a) and (b) are special cases of the First Fixed Point Theorem.

In what follows $t(x, x_1, \ldots, x_n)$ is a term in which $x, x_1, \ldots, x_n$ are precisely the variables that occur in the term. Recall that all realizable terms are built up only from variables (and parentheses).

Theorem F_1. [First Fixed Point Theorem] *For any realizable term* $t(x, x_1, \ldots, x_n)$ *there is a combinator* Γ *satisfying the condition* $\Gamma x_1 \ldots x_n = t(\Gamma, x_1, \ldots, x_n)$.

Remarks. This theorem may be a bit startling, since the satisfying condition for Γ involves Γ itself! This is related to self-reference or recursion, as will be seen in the next chapter.

Problem 5.

(1) Prove Theorem F_1. [There are two different proofs of this. One involves a fixed point of a combinator that realizes the term $t(x, x_1, \ldots, x_n)$. The other uses a combinator that realizes the term $t((xx), x_1, \ldots, x_n)$.]

(2) Now find solutions to (a) and (b) of the exercise stated before Theorem F_1.

Doubling Up

Theorem F_1 has the following double analogue:

Theorem $F_1 F_1$. [First Double Fixed Point Theorem] *For any two realizable terms* $t_1(x, x_1, \ldots, x_n)$ *and* $t_2(x, x_1, \ldots, x_n)$ *there are combinators* Γ_1, Γ_2 *such that:*
(1) $\Gamma_1 x_1 \ldots x_n = t_1(\Gamma_2, x_1, \ldots, x_n)$.
(2) $\Gamma_2 x_1 \ldots x_n = t_2(\Gamma_1, x_1, \ldots, x_n)$.

Problem 6. Prove Theorem $F_1 F_1$.

II. Partial Systems of Combinatory Logic

λ-I Combinators

We will now consider combinators A defined by some condition $A x_1 \ldots x_n = t$, where t is a term all of whose variables are among the variables $x_1 \ldots x_n$. We recall that if all of the variables $x_1 \ldots x_n$ occur in t, then A is called a λ-I combinator, or a non-cancellative combinator, while all other combinators (i.e. those for which one or more of the variables $x_1 \ldots x_n$ do not occur in t) are called cancellative combinators (e.g. K and KI), since they cancel out those of the variables $x_1 \ldots x_n$ that do not occur in t.

For any class $\mathcal{C}$ of combinators, a set $\mathcal{S}$ of elements of $\mathcal{C}$ is called a *basis* for $\mathcal{C}$ if all elements of $\mathcal{C}$ are derivable from elements of S. We have already shown that the class of all combinators has a *finite basis*, namely the two elements S and K. We now wish to show that the four combinators S, B, C and I form a basis for the class of all λ-I combinators (and hence that B, M, T and I also form a basis, since S, B, C and I are themselves derivable from B, M, T and I).

To this end, let us define a *nice term* to be a term built from the symbols S, B, C and I and variables. If a nice term has no variables, then it is a combinator, and will accordingly be called a *nice combinator*. Thus a nice combinator is a combinator built only from S, B, C and I.

We define an x-eliminate of a term as before and we must now show that if $\mathcal{W}$ is any nice term, then $\mathcal{W}$ has a nice x-eliminate, *provided that x actually occurs in $\mathcal{W}$!* Well, for this purpose we must redefine our set of recursive rules to produce the appropriate x-eliminate $\mathcal{W}_x$.

1. If $\mathcal{W} = x$, we take $\mathcal{W}_x$ to be I. [Thus $x_x = I$.]
2. For a compound term $\mathcal{W}\mathcal{V}$ in which x occurs, it occurs either in $\mathcal{W}$ or in $\mathcal{V}$ or in both.

   (a) If x occurs both in $\mathcal{W}$ and $\mathcal{V}$, we take $(\mathcal{W}\mathcal{V})_x$ to be $S\mathcal{W}_x\mathcal{V}_x$.
   (b) If x occurs in $\mathcal{W}$ but not in $\mathcal{V}$, we take $(\mathcal{W}\mathcal{V})_x$ to be $C\mathcal{W}_x\mathcal{V}$.
   (c) If x occurs in $\mathcal{V}$ but not in $\mathcal{W}$, then:

      (c_1) if $\mathcal{V}$ consists of x alone, we take $(\mathcal{W}\mathcal{V})_x$ to be $\mathcal{W}$. [Thus $(\mathcal{W}x)_x = \mathcal{W}$, if x does not occur in $\mathcal{W}$.]
      (c_2) if $\mathcal{V} \neq x$ (but x occurs in $\mathcal{V}$ and not in $\mathcal{W}$), we take $(\mathcal{W}\mathcal{V})_x$ to be $B\mathcal{W}\mathcal{V}_x$.

Note that this set of rules, unlike the previous set of rules for x-elimination, contains no rule for eliminating x when the original term you wish to eliminate x from is a term in which x doesn't occur. But this causes no problem in our current non-cancellative combinator situation. If our *original* term contains no occurrence of x, we would have no reason to even start off on an x-elimination, because we only eliminate variables from the list of arguments for the desired combinator, and if the original term contains no x, there would be no x in the argument list, so we wouldn't try to do x-elimination at all. So we are only concerned with doing an x-elimination either on an x standing alone, which is covered by Rule (1), or on a compound term $\mathcal{W}\mathcal{V}$ in which x occurs in either $\mathcal{W}$ or $\mathcal{V}$ or both. If we do find after eliminating x from such a compound term (and we are told by the rules as to what to do when x doesn't occur in one of the terms) that there are no longer any x's in the compound term that result, we just know that we have the final x-eliminate for the term (or subterm) we are working on.

Proposition 3. *For any nice term $\mathcal{W}$ in which x occurs, the term $\mathcal{W}_x$ is a nice x-eliminate of $\mathcal{W}$.*

Problem 7. Prove Proposition 3.

We leave it to the reader to show that for a nice term $\mathcal{W}$, if variables x and y occur in $\mathcal{W}$, then $(\mathcal{W}_x)_y$ is a nice y, x eliminate of $\mathcal{W}$, and more generally that if $x_1, \ldots, x_n$ occur in $\mathcal{W}$, then $\mathcal{W}_{x_n x_{n-1} \ldots x_1}$ is a nice $x_1, \ldots, x_n$ eliminate of $\mathcal{W}$, and so if $x_1, \ldots, x_n$ are all the variables of $\mathcal{W}$, then $\mathcal{W}_{x_n x_{n-1} \ldots x_1}$ is a nice combinator A satisfying the condition $A x_1 \ldots x_n = \mathcal{W}$. Since A is nice, then it is also a λ-I combinator (since S, B, C and I are all λ-I combinators, and for any two λ-I combinators $A_1 A_2$ is again a λ-I combinator).

Notice that this time we are proving Proposition 3 for every term built up from S, B, C, I *and* variables (we said *any* nice term). I pointed out earlier we also could have done this for terms built from S and K and variables with the earlier method of eliminates and the same proofs. But since terms built up only from variables are also terms built up from S, B, C, I and possibly variables, we have the following immediate consequence of Proposition 3:

Corollary. *For every* constant-free *term t and every list of possible arguments $x_1, x_2, \ldots, x_n$ for a combinator, if the variables occurring in the term t are $x_1, x_2, \ldots, x_n$ too, there is a combinator A in the class of variable-free terms built from S, B, C and I such that $A x_1 x_2 \ldots x_n = t$ (i.e. there is a non-cancellative combinator A that realizes the term t).*

This proves that S, B, C and I form a basis for the class of all λ-I combinators.

It is known that no *proper* subset of $\{S, B, C, I\}$ is a basis for the class of λ-I combinators, but J. B. Rosser found a curious two-element basis for the class of λ-I combinators, namely I and a combinator J defined by the condition:
$$J x y z w = x y (x w z).$$

How Rosser found that is a mystery to me! If the reader is interested, a derivation of B, T and M from J and I can be found in *To Mock a Mockingbird*, and probably in other books on combinatory logic.

B, T, I Combinators

The class of all combinators derivable from B, T and I has been studied by Rosser [1936] in connection with certain logical systems in which duplicative combinators like M, W, S, J have no place. In *To Mock a Mockingbird* I showed that these three combinators can be replaced by only two, namely

I and the combinator *G* (the goldfinch), which we recall is defined by the condition:

$$Gxyzw = xw(yz).$$

Whether this discovery was new or not, I honestly don't know. Now let us derive *B* and *T* from *G* and *I*.

Problem 8.

(a) First derive from *G* and *I* the combinator Q_3 (which, we recall, is defined as satisfying the condition $Q_3xyz = z(xy)$).

(b) Then derive the cardinal *C* from *G* and *I*. [$Cxyz = xzy$.]

(c) From *C* and *I*, we can obtain *T*. [Alternatively, we can get *T* from Q_3 and *I*].

(d) From *C* we can get the robin *R* [$Rxyz = yzx$], and then from *C, R* and Q_3, we can get the queer combinator *Q* [$Qxyz = y(xz)$], and then from *Q* and *C* we can obtain *B*.

Solutions to the Problems of Chapter 11

1. *t* is a term if and only if there is a finite sequence $t_1, \ldots, t_n = t$ such that for each $i \le n$, either t_i is a variable or constant, or there are numbers j_1 and j_2 both less than *i* such that $t_i = (t_{j_1}, t_{j_2})$.

2. We will show that no matter which rule applies to our simple term $\mathcal{W}$, then $\mathcal{W}_x$ will be an *x*-eliminate of $\mathcal{W}$.

   (i) Let us first consider the case that $\mathcal{W}$ is *x* itself. Then, by rule (1), $\mathcal{W}_x = I$. We must thus show that *I* is an *x*-eliminate of $\mathcal{W}$. Well, since there are no variables in the term *I*, then of course the variable *x* does not occur in *I*, and it is vacuously true that all variables in *I* (of course there are none) occur in $\mathcal{W}$. Also, $Ix = x$, and so *I* is an *x*-eliminate of $\mathcal{W}$.

   (ii) Now we will show that if $\mathcal{W}$ is any term in which *x* does not occur, then, by rule (2) $K\mathcal{W}$ is an *x*-eliminate of $\mathcal{W}$. Clearly, *x* does not occur in $K\mathcal{W}$. Secondly, the variables of $K\mathcal{W}$ are variables of $\mathcal{W}$, so that all variables of $K\mathcal{W}$ are variables of $\mathcal{W}$. Thirdly, $K\mathcal{W}x = \mathcal{W}$. Thus $K\mathcal{W}$ is indeed an *x*-eliminate of $\mathcal{W}$.

   (iii) Finally, we consider the composite term $\mathcal{W}\mathcal{V}$ in which *x* does occur, so that rule (3) applies. Suppose that $\mathcal{W}_x$ is an *x*-eliminate of $\mathcal{W}$, and that $\mathcal{V}_x$ is an *x*-eliminate of $\mathcal{V}$. We must show that $S\mathcal{W}_x\mathcal{V}_x$ is an

x-eliminate of $\mathcal{WV}$. Let us check the three conditions to see if $S\mathcal{W}_x\mathcal{V}_x$ is an x-eliminate of $\mathcal{WV}$:

(a) Since x does not occur in $\mathcal{W}_x$ (since $\mathcal{W}_x$ is an x-eliminate of $\mathcal{W}$) and x does not occur in $\mathcal{V}_x$), and x does not occur in S, it follows that x does not occur in $S\mathcal{W}_x\mathcal{V}_x$.

(b) Every variable y in $S\mathcal{W}_x\mathcal{V}_x$ must occur in $\mathcal{WV}$, because y either occurs in $\mathcal{W}_x$ or $\mathcal{V}_x$ or both, which means it occurs in $\mathcal{W}$ or $\mathcal{V}$ or both. So it occurs in the compound term $\mathcal{WV}$.

(c) Lastly, we must show that $S\mathcal{W}_x\mathcal{V}_x x = \mathcal{WV}$. Well,

$$S\mathcal{W}_x\mathcal{V}_x x = \mathcal{W}_x x(\mathcal{V}_x x).$$

Since $\mathcal{W}_x x = \mathcal{W}$ and $\mathcal{V}_x x = \mathcal{V}$, then $\mathcal{W}_x x(\mathcal{V}_x x) = \mathcal{WV}$. Thus $S\mathcal{W}_x\mathcal{V}_x x = \mathcal{WV}$.

By (a), (b), and (c), $S\mathcal{W}_x\mathcal{V}_x$ is an x-eliminate of $\mathcal{WV}$ (assuming $\mathcal{W}_x$ is an x-eliminate of $\mathcal{W}$ and assuming $\mathcal{V}_x$ is an x-eliminate of $\mathcal{V}$).

3. (a) The variable x does not occur in $\mathcal{W}$ (by hypothesis).
   (b) All variables of $\mathcal{W}$ obviously occur in $\mathcal{W}x$.
   (c) Obviously $\mathcal{W}x = \mathcal{W}x$.

4. It is clear that the statement (2) is true. Here is why:
   (a) The variable y does not occur in $(\mathcal{W}_x)_y$. The variable x does not occur in $\mathcal{W}_x$, hence not in $(\mathcal{W}_x)_y$, since all the variables in $(\mathcal{W}_x)_y$ occur in $\mathcal{W}_x$. Thus neither x nor y occurs in $\mathcal{W}_{x,y}$.
   (b) All variables of $\mathcal{W}_x$ occur in $\mathcal{W}$, and all variables in $(\mathcal{W}_x)_y$ occur in $\mathcal{W}_x$. Hence all variables of $\mathcal{W}_{x,y}$ occur in $\mathcal{W}$.
   (c) $(\mathcal{W}_x)_y y = \mathcal{W}_x$, so that $(\mathcal{W}_x)_y yx = \mathcal{W}_x x = \mathcal{W}$.
   By (a), (b), and (c), $(\mathcal{W}_x)_y$ is an y, x eliminate of $\mathcal{W}$.

5. For the first proof, let A be a combinator that realizes the term t. Thus

$$Axx_1\ldots x_n = t(x, x_1, \ldots, x_n).$$

We let Γ be a fixed point of A, so that

$$\Gamma x_1\ldots x_n = A\Gamma x_1\ldots x_n = t(\Gamma, x_1, \ldots, x_n).$$

Thus $\Gamma x_1\ldots x_n = t(\Gamma, x_1, \ldots, x_n)$.

For the second proof, suppose A realizes the term $t(xx, x_1, \ldots, x_n)$, i.e. for all $x, x_1, \ldots, x_n$

$$Axx_1\ldots x_n = t(xx, x_1, \ldots, x_n).$$

If we take A for x, we obtain

$$AAx_1 \ldots x_n = t(AA, x_1, \ldots, x_n).$$

We thus take AA to be Γ!

Note to the reader: I believe that virtually all fixed points and recursion theorems in recursion theory and in combinatory logic, as well as the construction of Gödel self-referential sentences, are simply elaborate variations on this trick of the above second proof. In all cases, one has an equation containing Ax of the left side of an equal sign, and xx on the right side. Taking A for x, we end up with AA on both sides.

Now here is the solution of the Exercise:

(a) Since $Qzxy = x(zy)$, we take Γ to be a fixed point of Q, so that

$$\Gamma xy = Q\Gamma xy = x(\Gamma y).$$

Thus $\Gamma xy = x(\Gamma y)$.

(b) Since $Czxy = zyx$, we take Γ to be a fixed point of C, so that

$$\Gamma xy = C\Gamma xy = \Gamma yx.$$

Thus $\Gamma xy = \Gamma yx$.

6. Again, there are two different ways of proving this, and both are of interest.

For the first proof, we recall that by a *cross point* of a pair of compatible points (A, B) is meant a pair of elements (x, y) such that $Ax = y$ and $By = x$. Since we are working with a complete system, the hypothesis of Problem 6, Chapter 8 (that conditions C_1 and C_2 both hold) is met, so that every pair of terms has a cross point (in the language of that problem, any two combinators are compatible).

Now let A_1 realize t_1 and let A_2 realize t_2. Thus

(3) $A_1xx_1 \ldots x_n = t_1(x, x_1, \ldots, x_n).$
(4) $A_2xx_1 \ldots x_n = t_2(x, x_1, \ldots, x_n).$

We now let (Γ_1, Γ_2) be a cross point of (A_2, A_1) [not of (A_1, A_2) but of (A_2, A_1)]. Thus, $A_2\Gamma_1 = \Gamma_2$ and $A_1\Gamma_2 = \Gamma_1$. Then

(1') $\Gamma_1x_1 \ldots x_n = A_1\Gamma_2x_1 \ldots x_n = t_1(\Gamma_2, x_1, \ldots, x_n).$
(2') $\Gamma_2x_1 \ldots x_n = A_2\Gamma_1x_1 \ldots x_n = t_2(\Gamma_1, x_1, \ldots, x_n).$

For the second proof, given terms $t_1(x, x_1, \ldots, x_n)$ and $t_2(x, x_1, \ldots, x_n)$, let A_1 and A_2 be combinators satisfying the following conditions:

(1) $A_1 xyx_1 \ldots x_n = t_1(xyx, x_1, \ldots, x_n)$.

(2) $A_2 xyx_1 \ldots x_n = t_2(xyx, x_1, \ldots, x_n)$.

Then,

(1') $A_1 A_2 A_1 x_1 \ldots x_n = t_1(A_2 A_1 A_2, x_1, \ldots, x_n)$.

(2') $A_2 A_1 A_2 x_1 \ldots x_n = t_2(A_1 A_2 A_1, x_1, \ldots, x_n)$.

We thus take $\Gamma_1 = A_1 A_2 A_1$ and $\Gamma_2 = A_2 A_1 A_2$.

7. We are given that $\mathcal{W}$ is a nice term and that x occurs in $\mathcal{W}$. At each stage of the construction of $\mathcal{W}$, no new constant was introduced other than S, B, C or I, so that $\mathcal{W}_x$ is certainly nice.

   We leave it to the reader, using term induction, to prove that x does not occur in $\mathcal{W}_x$ and that all variables in $\mathcal{W}_x$ occur in $\mathcal{W}$. It remains to show that $\mathcal{W}_x x = \mathcal{W}$. We do this by induction.

   1   If $\mathcal{W} = x$, then $\mathcal{W}_x$ is I and $\mathcal{W}_x x = Ix = I\mathcal{W} = \mathcal{W}$. So $\mathcal{W}_x x = \mathcal{W}$. It remains to show the statement for when we have a compound term $\mathcal{W}\mathcal{V}$.

   2(a) Suppose that x is in both $\mathcal{W}$ and $\mathcal{V}$, and that $\mathcal{W}_x x = \mathcal{W}$ and $\mathcal{V}_x x = \mathcal{V}$. We are to show that $(\mathcal{W}\mathcal{V})_x x = \mathcal{W}\mathcal{V}$. Well,

   $$(\mathcal{W}\mathcal{V})_x = S\mathcal{W}_x \mathcal{V}_x,$$

   so that

   $$(\mathcal{W}\mathcal{V})_x x = S\mathcal{W}_x \mathcal{V}_x x = \mathcal{W}_x x(\mathcal{V}_x x) = \mathcal{W}\mathcal{V}.$$

   2(b) For the case when x occurs in $\mathcal{W}$ but not in $\mathcal{V}$, we assume by the induction hypothesis that $\mathcal{W}_x x = \mathcal{W}$. Now, in this case,

   $$(\mathcal{W}\mathcal{V})_x = C\mathcal{W}_x \mathcal{V},$$

   so that

   $$(\mathcal{W}\mathcal{V})_x x = (C\mathcal{W}_x \mathcal{V})x = \mathcal{W}_x x\mathcal{V} = \mathcal{W}\mathcal{V}.$$

   2(c) Now for the case that x is in $\mathcal{V}$ but not in $\mathcal{W}$.

   (c_1) Suppose $\mathcal{V} = x$. Then $(\mathcal{W}\mathcal{V})_x = \mathcal{W}$, so that

   $$(\mathcal{W}\mathcal{V})_x x = \mathcal{W}x = \mathcal{W}\mathcal{V}.$$

   (c_2) In the case that $\mathcal{V}_x \neq x$ (and when x does not occur in $\mathcal{W}$), we assume that $\mathcal{V}_x x = \mathcal{V}$ and we must show that $(\mathcal{W}\mathcal{V})_x x = \mathcal{W}\mathcal{V}$. Well, in this case $(\mathcal{W}\mathcal{V})_x = B\mathcal{W}\mathcal{V}_x$, so that

   $$(\mathcal{W}\mathcal{V})_x x = B\mathcal{W}\mathcal{V}_x x = \mathcal{W}(\mathcal{V}_x x) = \mathcal{W}\mathcal{V}.$$

This concludes the proof.

8. (a) We take Q_3 to be GI. Then $GIxyz = Iz(xy) = z(xy) = Q_3xyz$.

   (b) We take C to be $GGII$. Then $GGIIx = Gx(II) = GxI$, so that

$$GGIIxyz = GxIyz = xz(Iy) = xzy = Cxyz.$$

   (c) We can take T to be CI. Then $CIxy = Iyx = yx = Txy$. Alternatively, we can take T to be Q_3I. Then

$$Q_3Ixy = y(Ix) = yx = Txy.$$

   (d) We recall that CC is the robin R, since

$$CCxyz = Cyxz = yzx = Rxyz.$$

   We now take Q to be GRQ_3. Then

$$GRQ_3xyz = Ry(Q_3x)z = Q_3xzy = y(xz) = Qxyz.$$

   Finally, we take B to be CQ. Then

$$CQxyz = Qyxz = x(yz) = Bxyz.$$

Chapter 12

Combinators, Recursion
and the Undecidable

In this chapter, we work with the complete system: all combinators derivable from S and K. We shall now see how combinatory logic is related to propositional logic, recursion theory, and the undecidable.

Logical Combinators

We now want two combinators t and f to stand for truth and falsity respectively. Several workable choices are possible, and the one we take is due to Henk Barendregt [1985]. We take t to be the kestrel K, and for f we take the combinator KI, so that in what follows t is an abbreviation for K, while f is an abbreviation for KI. We shall call t and f *propositional combinators*. We note that for any combinators x and y, whether propositional combinators or not, $txy = x$ and $fxy = y$, which turns out to be a technical advantage.

There are only two propositional combinators t and f (i.e. K and KI). We now use the letters p, q, r, s as standing for arbitrary *propositional combinators*, rather than propositions, and we call p *true* if $p = t$, and *false* if $p = f$. We define the *negation* $\sim p$ of p by the usual conditions: $\sim t = f$ and $\sim f = t$; we define the *conjunction* $p \wedge q$ by the truth-table conditions: $t \wedge t = t$, $t \wedge f = f$, $f \wedge t = f$, $f \wedge f = f$. We define the *disjunction* $p \vee q$, the *conditional* $p \supset q$, and the *bi-conditional* $p \equiv q$ similarly. We now want combinators that can realize these operations.

Problem 1. Find combinators N (negation combinator), c (conjunction combinator), d (disjunction combinator), i (implication combinator), and e

(equivalence or bi-conditional combinator), satisfying the following conditions:

(a) $Np = \sim p$,

(b) $cpq = p \wedge q$,

(c) $dpq = p \vee q$,

(d) $ipq = p \supset q$,

(e) $epq = p \equiv q$.

We now see that by combining the above combinators in various ways, we have combinators that can do propositional logic, in the sense that they can compute compound truth tables.

Arithmetic Combinators

We shall now see how the arithmetic of the natural numbers can be embedded in combinatory logic.

Each natural number n is represented by a combinator denoted $\overline{n}$ (not to be confused with the Peano numeral designating n). There are several schemes of doing this, one of which is due to Church [1941]. The scheme we will use is again due to Henk Barendregt [1985]. Here the vireo V plays a major role $[Vxyz = zxy]$. We let σ (read as "sigma") be the *successor combinator* Vf (which is $V(KI)$). We take $\overline{0}$ to be the identity combinator I, take $\overline{1}$ to be $\sigma\overline{0}$; take $\overline{2}$ to be $\sigma\overline{1}, \ldots$, take $\overline{n+1} = \sigma\overline{n}, \ldots$. The first thing that needs to be shown is that the combinators $\overline{0}, \overline{1}, \overline{2}, \ldots$, are all distinct.

Problem 2. Show that if $m \neq n$, then $\overline{m} \neq \overline{n}$. [**Hint:** First show that for every n, $\overline{0} \neq \overline{n^+}$, where n^+ is an abbreviation for $n + 1$. Then show that if $\overline{n^+} = \overline{m^+}$, then $\overline{n} = \overline{m}$. Finally, show, for all m and n and all positive k, that $\overline{n} \neq \overline{n+k}$.]

The combinators $\overline{0}, \overline{1}, \overline{2}, \ldots$ are called *numerical combinators*. A combinator A is called an *arithmetical combinator of type* 1 if, for every n, there is some m such that $A\overline{n} = \overline{m}$. We call A an arithmetical combinator of type 2 if, for every n and m, there is some number p such that $A\overline{n}\,\overline{m} = \overline{p}$. In general, A is called an arithmetical combinator of type n if, for all numbers $x_1, \ldots, x_n$, there is a number y such that $A\overline{x_1} \ldots \overline{x_n} = \overline{y}$.

We now wish to show that there is an *addition combinator*, which we will denote $\oplus$, such that $\oplus\overline{m}\,\overline{n} = \overline{m+n}$, and a *multiplication combinator*, which we will denote $\otimes$, such that $\otimes \overline{m}\,\overline{n} = \overline{m \times n}$, and an *exponential combinator* $\text{\textcircled{E}}$ such that $\text{\textcircled{E}}\,\overline{m}\,\overline{n} = \overline{m^n}$.

We need some items in preparation for all this:

For any *positive* number n, by its *predecessor* is meant the number $n - 1$. Thus, for any number n, the predecessor of n^+ is n. We now need a combinator P that calculates predecessors, in the sense that $P\overline{n^+} = \overline{n}$ for all n.

Problem 3. Find such a combinator P.

The Zero Tester

We have vital need of a combinator Z, called the *zero tester*, such that $Z\overline{0} = t$ and $Z\overline{n} = f$ for any positive n.

Problem 4. Find such a combinator Z.

Problem 5. Does there exist a combinator A such that $A\overline{0}xy = x$, but $A\overline{n}xy = y$, for $\overline{n}$ positive?

We now consider some recursion properties.

Problem 6. Prove that for any combinator A and any number k, there is a combinator Γ satisfying the following two conditions:

$$C_1: \quad \Gamma\overline{0} = \overline{k}.$$
$$C_2: \quad \Gamma\overline{n^+} = A(\Gamma\overline{n}).$$

Hints:
1. C_2 is equivalent to the following condition: For any positive n,

$$\Gamma\overline{n} = A(\Gamma(P\overline{n})).$$

2. Use the zero tester Z as in Problem 5.
3. Use the fixed point theorem.

Problem 7. Now show that there are combinators $\oplus$, $\otimes$ and $\text{\textcircled{E}}$ satisfying the following conditions:
(1) $\oplus\,\overline{m}\,\overline{n} = \overline{m + n}$.
(2) $\otimes\,\overline{m}\,\overline{n} = \overline{m \times n}$.
(3) $\text{\textcircled{E}}\,\overline{m}\,\overline{n} = \overline{m^n}$.

The Recursion Property for Combinatory Logic, in its most general form, is that for any terms $f(y_1, \ldots, y_n)$ and $g(x, z, y_1, \ldots, y_n)$, there is a combinator Γ such that for all numerals $\overline{x}, \overline{y_1}, \ldots, \overline{y_n}$, the following two conditions hold:

(1) $\Gamma \overline{0} \overline{y_1} \ldots \overline{y_n} = f(\overline{y_1}, \ldots, \overline{y_n})$.

(2) $\Gamma \overline{x'} \overline{y_1} \ldots \overline{y_n} = g((P\overline{x'})(\Gamma(P\overline{x'})\overline{y_1} \ldots \overline{y_n}), \overline{y_1}, \ldots, \overline{y_n})$.

A combinator Γ that does this exists by the fixed point theorem, and is defined by the condition:

$$\Gamma x y_1 \ldots y_n = Zx f(y_1, \ldots, y_n) g((Px), (\Gamma(Px)y_1 \ldots y_n), y_1, \ldots, y_n).$$

This brilliant idea was due to Alan Turing!

I.   Preparation for the Finale

Property Combinators

By a *property combinator* is meant a combinator such that for every number n, $\Gamma \overline{n} = t$ or $\Gamma \overline{n} = f$.

A set A of (natural) numbers is said to be (*combinatorially*) *computable* if there is a property combinator Γ such that $\Gamma \overline{n} = t$ for every n in A, and $\Gamma \overline{n} = f$ for every n that is not in A. Such a combinator Γ is said to *compute* A.

We might remark that the important thing about combinatorial computability is that a set A is combinatorially computable if and only if it is recursive.

Problem 8. Show that the set E of even numbers is computable. [**Hint**: The property of being even is the one and only property satisfying the following conditions: (1) 0 is even; (2) for positive n, n is even if and only if its predecessor is not even. Now use the fixed point theorem.]

Problem 9. Suppose Γ computes A. Does it follow that $N\Gamma$ computes the complement $\overline{A}$ of A? [N is the negation combinator Vft.]

Relational Combinators

By a relational combinator of degree n is meant a combinator A such that for all numbers $k_1, \ldots, k_n$, either $A\overline{k_1} \ldots \overline{k_n} = t$ or $A\overline{k_1} \ldots \overline{k_n} = f$, and such a combinator A is said to compute the set of all n-tuples $(k_1, \ldots, k_n)$ such that $A\overline{k_1} \ldots \overline{k_n} = t$. Thus for any relation $R(x_1, \ldots, x_n)$, we say that A computes R to mean that for all n-tuples $(k_1, \ldots, k_n)$, if $R(k_1, \ldots, k_n)$ holds, then $A\overline{k_1} \ldots \overline{k_n} = t$, and if $R(k_1, \ldots, k_n)$ doesn't hold, then $A\overline{k_1} \ldots \overline{k_n} = f$.

Problem 10. Show that there is a relational combinator g that computes the relations $x > y$ (x is greater than y). [**Hint**: This relation is uniquely determined by the following conditions:

1. If $x = 0$, then $x > y$ is false.
2. If $x \neq 0$ and $y = 0$, then $x > y$ is true.

   If $x \neq 0$ and $y \neq 0$, then $x > y$ is true if and only if $x - 1 > y - 1$ is true.]

Functional Combinators

Consider a function $f(x)$ of one argument, i.e. an operation that assigns to each number n a number devoted by $f(n)$. A combinator A will be said to *realize* the function $f(x)$ if, for every number n, the condition $A\overline{n} = \overline{f(n)}$ holds.

Problem 11. Show that if functions $f(x)$ and $g(x)$ are both realizable, so is the function $f(g(x))$, i.e. the operation that assigns to each number n the number $f(g(n))$.

Problem 12. Suppose $R(x, y)$ is computable and that $f(x)$ and $g(x)$ are both realizable. Show that the following are computable:
(a) The relation $R(f(x), y)$.
(b) The relation $R(x, g(y))$.
(c) The relation $R(f(x), g(y))$.

The Minimization Principle

Consider a relational combinator A of degree 2 such that for every number n, there is at least one number m such that $A\overline{n}\,\overline{m} = t$. Such a combinator is sometimes called *regular*. If A is regular, then, for every n, there must be a smallest number k such that $A\overline{n}\overline{k} = t$. The minimization principle is that for every regular combinator A, there is a combinator A', called a *minimizer* of A, such that for every number n, $A'\overline{n} = \overline{k}$, where k is the smallest number for which $A\overline{n}\overline{k} = t$.

We seek to prove the minimization principle. The following preliminary problem will be most helpful:

Problem 13. Show that for any regular relational combinator A of degree 2, there is a combinator A_1 such that for all numbers n and m,
(1) If $A\overline{n}\,\overline{m} = t$, then $A_1\overline{n}\,\overline{m} = \overline{m}$.
(2) If $A\overline{n}\,\overline{m} = f$, then $A_1\overline{n}\,\overline{m} = \overline{A_1\overline{n}\overline{m} + 1}$.

Problem 14. Now prove the Minimization Principle: for every regular relational combinator A of degree 2, there is a combinator A', called a *minimizer* of A, such that for every number n, $A'\overline{n} = \overline{k}$, where k is the smallest number for which $A\overline{n}\overline{k} = t$. [**Hint:** Use the A_1 of Problem 13 and the cardinal C $[Cxyz = xzy]$.]

The Length Measurer

By the length of a number n we shall mean the number of digits in n when n is in ordinary base 10 notation. The numbers from 0 to 9 have length 1; those from 10 to 99 have length 2, those from 100 to 999 have length 3, etc. Actually the length of n is the smallest number k such that $10^k > n$.

We now want a combinator L' that measures the length of any number, that is, we want such that $L'\overline{n} = \overline{k}$, where k is the length of n.

Problem 15. Prove that there is a length measurer L'.

Concatenation to the Base 10

For any numbers n and m, by $n * m$ we shall now mean the number which, when written in ordinary base 10 notation, consists of n (in base 10 notation) followed by m (in base 10 notation). For example

$$47 * 386 = 47386.$$

We now want a combinator ⊛ that realizes this operation of concatenation $*$.

Problem 16. Show that there is a combinator $\circledast$ such that for every n and m, $\circledast\,\overline{n}\,\overline{m} = \overline{n * m}$. [**Hint:** $x * y = (x \times 10^k) + y$, where k is the length of y. For example,

$$26 * 587 = 26000 + 587 = (26 \times 10^3) + 587,$$

and 3 is the length of 587.]

II.   The Grand Problem

By an S, K term we shall mean a term without variables, built from just the constants S and K. We use letters X, Y, Z to stand for unspecified S, K terms. In what follows "term" will mean S, K term.

Of course, distinct terms might designate the same combinator. For example, KKK designates the same combinator as KKI. For both designate the kestrel K, or as we say the equation $KKK = KKI$ holds, or is *true*. An equation is true if and only if its being so is a consequence of the given conditions (which can be called the *axioms for combinatory logic*) — $SXYZ = XZ(YZ)$ and $KXY = X$ — and by using the inference rules of identity, namely $X = X$ and if $X = Y$, then $XZ = YZ$ and $ZX = ZY$; also, the additional rules of equality: if $X = Y$ then $Y = X$; if $X = Y$ and $Y = Z$ then $X = Z$.

The grand question now is this: Given two terms X and Y, is there a systematic way of deciding whether or not the equation $X = Y$ holds? This can be reduced to a question of deciding whether a given number belongs to a certain set of numbers. This translation uses the device of Gödel numbering. How this translation is effected is the first thing we shall now consider.

All sentences (frequently called equations here) are built from the following five symbols:

$$\begin{array}{ccccc} S & K & (&) & = \\ 1 & 2 & 3 & 4 & 5 \end{array}$$

In this section all such sentences will be equations of the form

$$S, K \text{ term } = S, K \text{ term}.$$

Under each symbol I have written its Gödel number. The Gödel number of a compound expression (a complex term, or a sentence) is the number obtained by replacing S by 1, K by 2, ..., $=$ by 5, and reading the result as a number in base 10. For example, the Gödel number of $KS (=$ is 2135).

We let $\mathcal{T}$ be the set of true sentences, and let $\mathcal{T}_0$ be the set of the Gödel numbers of the true sentences. The question now is whether the set $\mathcal{T}_0$ is combinatorially computable. Is there a combinator Γ such that for every n, $\Gamma \bar{n} = t$ if $n \in \mathcal{T}_0$, and $\Gamma \bar{n} = f$ if $n \notin \mathcal{T}_0$?

The question is of staggering importance, since any *formal* mathematical question can be reduced to whether a certain number belongs to $\mathcal{T}_0$. The question is thus equivalent to the question of whether there can be a universal computer that can settle all formal mathematical questions.

As the reader has doubtless guessed, the answer is *no*, which we now set out to prove.

Gödel Numerals

Let me first point out that there are two different schemes for constructing terms, which involve two different ways of introducing parentheses. One scheme consists of building terms according to the following two rules:

(1) S and K are terms.

(2) If X and Y are terms, so is (XY).

The second scheme replaces (2) by:

(2′) If X and Y are terms, so is $(X)(Y)$.

We shall continue to use the first scheme.

Now, for any number n, the numeral $\bar{n}$, like any other term, has a Gödel number. We let $n^{\#}$ be the Gödel number of the numeral $\bar{n}$. For example, $\bar{0}$ is the combinator I, which in terms of S and K is $((SK)K)$, whose Gödel number is 3312424. Thus $0^{\#} = 3312424$.

As for $1^{\#}$, it is the Gödel number of the numeral $\bar{1}$, and $\bar{1} = (\sigma\bar{0})$, where σ is the combinator Vf, and Vf, when reduced to S and K, is a horribly long expression, and consequently has an unpleasantly long Gödel number, which I will abbreviate by the letter "s". Thus, in what follows, s is the Gödel number of (the term whose abbreviation is) σ, and so $1^{\#}$ is the Gödel number of $(\sigma\bar{0})$, which is $3 * s * 0^{\#} * 4$.

Next, $2^{\#}$ uses the Gödel number of $(\sigma\bar{1})$, and the Gödel number of that term is $3 * s * 1^{\#} * 4$. Similarly $3^{\#} = 3 * s * 2^{\#} * 4$. And so on $\ldots$.

We now need to show that the function $f(n) = n^{\#}$ is combinatorially realizable, i.e. that there is a combinator δ such that $\delta\bar{n} = \overline{n^{\#}}$ holds (for every n).

Problem 17. Show that there is such a combinator δ.

Normalization

For any expression X, by $\lceil X \rceil$ is meant the *numeral* that designates the Gödel number of X; we might call $\lceil X \rceil$ the *Gödel numeral* of X. By the *norm* of X is meant $X\lceil X \rceil$, i.e. X followed by the Gödel numeral of X. If n is the Gödel number of X, then $n^{\#}$ is the Gödel number of $\lceil X \rceil$, so $n * n^{\#}$ is the Gödel number of $X\lceil X \rceil$, i.e. of the norm of X.

We now want a combinator Δ, called a *normalizer*, such that $\Delta\bar{n} = \overline{n * n^{\#}}$ holds, for all n.

Problem 18. Exhibit such a combinator Δ.

Problem 19. Which, if either, of the following statements is true?
(a) $\Delta\lceil X \rceil = X\lceil X \rceil$
(b) $\Delta\lceil X \rceil = \lceil X\lceil X \rceil \rceil$

The normalizer can do some amazing things, as you will see!

The Second Fixed Point Principle

We say that a term X *designates* a number n if the equation $X = \bar{n}$ holds. Obviously one term that designates n is the numeral $\bar{n}$, but there are many others. For example, for $n = 8$, the terms $\oplus\bar{4}\,\bar{4}$, $\oplus\bar{5}\,\bar{3}$, $\oplus\bar{2}\,\bar{6}$ all designate 8. Indeed there are infinitely many terms that designate $\bar{n}$, e.g. $I\bar{n}$, $I(I\bar{n})$, $I(I(I\bar{n}))$, etc.

We call a term an *arithmetic term* if it designates some number. Every numeral is an arithmetic term, but not every arithmetic term is a numeral.

It is impossible for any numeral to designate its own Gödel number, because the Gödel number of $\bar{n}$ is larger than n ($n^{\#} > n$). However, this does not mean that no *arithmetic* term can designate its own Gödel number. In fact, there *is* an arithmetic term that designates its own Gödel number! Also there is one that designates twice its Gödel number; one that designates eight times its Gödel number plus 15 — indeed, for *any* realizable function $f(x)$, there is an arithmetic term X which designates $f(n)$, where n is the Gödel number of X! This is immediate from the second fixed point principle, which is that for any combinator A, there is a term X such that $A\lceil X \rceil = X$ holds. We now wish to prove this principle.

Problem 20. Prove the Second Fixed Point Theorem: show that for any combinator A, there is a term X such that $A\lceil X \rceil = X$ holds.

Problem 21. Prove that for any realizable function $f(x)$ there is an (arithmetic term) X that designates $f(n)$, where n is the Gödel number of X. [**Hint:** Use the second fixed point theorem.]

Representability

We shall say that a combinator Γ *represents* a number set A if for every number n, the equation $\Gamma \overline{n} = t$ holds iff $n \in A$. Thus Γ represents the set of all n such that $\Gamma \overline{n} = t$ holds. (Note that the combinatorial computability of a number set A by a combinator Γ that we worked with earlier requires more than representability, for in the case of combinatorial computability if $n \notin A$ then it must be the case that $\Gamma \overline{n} = f$, while for representability, it is only necessary that $\Gamma \overline{n} \neq t$.)

Problem 22. Suppose Γ computes A. Which, if either, of the following statements are true?
(1) A is representable.
(2) The complement $\overline{A}$ of A is representable.

Recall that for any function $f(x)$ and set A, by $f^{-1}(A)$ is meant the set of all numbers n such that $f(n) \in A$. Thus $n \in f^{-1}(A)$ iff $f(n) \in A$.

Problem 23. Prove that if the function $f(x)$ is realizable, and if A is representable, then $f^{-1}(A)$ is representable.

Gödel Sentences

We shall call a sentence (equation) X a *Gödel sentence* for a number set A if X is true if and only if the Gödel number of X is in A. We now aim to prove that if A is representable, then there is a Gödel sentence for A.

Problem 24. Suppose A is representable. Prove that the set of all n such that $(n * 52) \in A$ is representable. [The significance of $n * 52$ is that if n is the Gödel number of X, then $n * 52$ is the Gödel number of $X = t$.]

Problem 25. Now prove that for any representable set A, there is a Gödel sentence for A.

Now we have all the key pieces to prove that the set $\mathcal{T}_0$ of Gödel numbers of the true sentences is not computable.

Problem 26. Prove that $\mathcal{T}_0$ is not computable.

Discussion

We have just seen that the *complement* of the set $\mathcal{T}_0$ is not representable. What about the set $\mathcal{T}_0$? Is it representable? Yes, it is. As suggested earlier, combinatory logic can be formalized. The following is an axiom system for the full combinatory logic:

Symbols: S, K, $(,)$, $=$

Terms: As previously defined recursively

Sentences (or Equations): Expressions of the form $X = Y$, where X and Y are terms.

Axiom Schemes: For any terms X, Y, Z:
(1) $SXYZ = XZ(YZ)$
(2) $KXY = X$
(3) $X = X$.

Inference Rules

R_1 : From $X = Y$ to infer $Y = X$.
R_2 : From $X = Y$ to infer $XZ = YZ$.
R_3 : From $X = Y$ to infer $ZX = ZY$.
R_4 : From $X = Y$ and $Y = Z$ to infer $X = Z$.

It is relatively easy to construct an elementary formal system in which one can successively represent the set of terms, sentences, and provable sentences. Thus the set $\mathcal{T}$ is formally representable, from which it easily follows that the set T_0 is recursively enumerable. And it is well known that recursive enumerability is the same thing as representability in combinatory logic. Thus the set T_0 is indeed representable.

The set T_0 is another example of a recursively enumerable set that is not recursive.

As previously mentioned, any formal mathematical question can be reduced to a question of whether a certain number is in T_0, that is, one can associate with each formal mathematical question a number n such that $n \in T_0$ iff the question has an affirmative answer. Since there is no purely mechanical method of determining which numbers are in T_0, there is then no purely mechanical method of determining which formal mathematical statements are true and which are false. Mathematics requires intelligence and ingenuity. Any attempt to fully mechanize mathematics is doomed to failure. In the prophetic words of Emil Post [1944], "Mathematics is and must remain essentially creative." Or, in the witty words of Paul Rosenbloom: "Man can never eliminate the necessity of using his own intelligence, regardless of how cleverly he tries."

Solutions to the Problems of Chapter 12

1. (a) We need a negation combinator N satisfying the condition $Nx = xft$. Well, we can take N to be Vft, where V is the vireo $[Vxyz = zxy]$. Thus $Nx = Vftx = xft$ as desired. Then $Nt = tft = f$ (since $t = K$, or because $txy = x$, for all x and y), and $Nf = fft = t$ (since $f = KI$, or because $fxy = y$, for all x and y). Thus $Np =\sim p$, where p is either t or f.

   (b) This time we want the conjunction combinator c to be such that $cxy = xyf$. We take c to be Rf, where R is the Robin $[Rxyz = yzx]$. Thus $cxy = Rfxy = xyf$, as desired. Then (using the fact that for all x and y, $txy = x$ and $fxy = f$):

   (1) $ctt = ttf = t$,

   (2) $ctf = tff = f$,

   (3) $cft = ftf = f$,

   (4) $cff = fff = f$.

   (c) Now we want the disjunction combinator d to be such that

$$dxy = xty.$$

For this purpose, we take d to be Tt, where T is the thrush $[Txy = yx]$. Then $dxy = xty$ (verify!). And $dpq = p \lor q$ (verify all four cases!).

(d) Take the implication combinator i to be Rt, where R is the robin $[Rxyz = yzx]$. Then $ixy = xyt$, because $ixy = Rtxy = xyt$. Then $ipq = p \supset q$ (verify!).

(e) We want the equivalence combinator e to be such that

$$exy = xy(Ny).$$

For this we may take e to be CSN:

$$CSNxy = SxNy = xy(Ny).$$

Then epq has the same values as $p \equiv q$. Verify, recalling that we have shown above that $Nt = f$, that $Nf = t$, and that, for all x and y, $txy = x$ and $fxy = f$.

2. Suppose $\overline{0} = \overline{n^+}$ for some n. Thus $I = \sigma\overline{n} = Vf\overline{n}$. Consequently, $IK = Vf\overline{n}K = Kf\overline{n} = f$. Thus $IK = KI$, so that $K = KI$ contrary to Problem 19 of Chapter 8.

Next, we must show that if $\overline{n} \neq \overline{m}$, then $\overline{n^+} \neq \overline{m^+}$. To do so, we instead show the equivalent fact that if $\overline{n^+} = \overline{m^+}$ then $\overline{n} = \overline{m}$.

Well, suppose that $\overline{n^+} = \overline{m^+}$. Then $\sigma\overline{n} = \sigma\overline{m}$, or $Vf\overline{n} = Vf\overline{m}$, so that $Vf\overline{n}f = Vf\overline{m}f$. Applying V here, we see that $ff\overline{n} = ff\overline{m}$, which implies that $\overline{n} = \overline{m}$ [again since it is always true that $fxy = y$]. This proves that if $\overline{n} \neq \overline{m}$, then $\overline{n+1} \neq \overline{m+1}$.

Now, for any positive k, we saw at the beginning of this solution that $\overline{0} \neq \overline{k}$. After what we have just proved, we now see as well that it follows from $\overline{0} \neq \overline{k}$ that $\overline{1} \neq \overline{1+k}$, $\overline{2} \neq \overline{2+k}$, ..., $\overline{n} \neq \overline{n+k}$, ... This proves that if $m \neq n$, then $\overline{m} \neq \overline{n}$ (because if $m \neq n$ then for some positive k, $m = n + k$ or $n = m + k$).

3. Since $\overline{n^+} = \sigma\overline{n}$, we want a combinator P such that $P(\sigma\overline{n}) = \overline{n}$. We take P to be Tf, where T is the thrush $[Txy = yx]$. Then

$$P(\sigma\overline{n}) = Tf(\sigma\overline{n}) = \sigma\overline{n}f = Vf\overline{n}f = ff\overline{n} = \overline{n}.$$

Voila!

4. We take Z to be Tt, where T is the thrush $[Txy = yx]$. Recall that $\overline{0} = I$ and $\sigma = Vf$.

(1) $Z\overline{0} = Tt\overline{0} = \overline{0}t = It = t$. Thus $Z\overline{0} = t$.

(2) $Z\overline{n^+} = Tt\overline{n^+} = \overline{n^+}t = \sigma\overline{n}t = Vf\overline{n}t = tf\overline{n} = f$. Thus $Z\overline{n^+} = f$.

5. Yes, the zero tester Z is such an A.

$$Z\overline{0}xy = txy = x.$$
$$Z\overline{n^+}xy = fxy = y.$$

6. Conditions C_1 and C_2 are respectively equivalent to:

C_1': $\Gamma x = \overline{k}$ if $x = \overline{0}$.

C_2': $\Gamma x = A(\Gamma(Px))$ if $x = \overline{n}$, for n positive

Now, by Problem 5, $Zx\overline{k}(A(\Gamma(Px)))$ is $\overline{k}$ if $x = \overline{0}$, and is $A(\Gamma(Px))$ if $x = \overline{n}$, for n positive. We thus want a combinator Γ which satisfies the condition:
$$\Gamma x = Zx\overline{k}(A(\Gamma(Px))).$$

Well, such a combinator Γ exists by the fixed point theorem. Specifically, we can take Γ to be a fixed point of a combinator θ satisfying the condition: $\theta yx = Zx\overline{k}(A(y(Px)))$. Then $\Gamma x = \theta\Gamma x = Zx\overline{k}(A(\Gamma(Px)))$.

7. (1) The addition combinator $\oplus$ is uniquely determined by the following two conditions:

    a. $n + 0 = n$.

    b. $n + m^+ = (n + m)^+$.

We therefore want a combinator $\oplus$ such that:

    a'. $\oplus\overline{n}\overline{0} = \overline{n}$.

    b'. $\oplus\overline{n}\overline{m^+} = \sigma(\oplus\overline{n}\,\overline{m})$.

We thus want $\oplus$ to be such that, for any n and m, whether 0 or positive, the following holds: $\oplus\overline{n}\,\overline{m} = Z\overline{m}\,\overline{n}(\sigma(\oplus(\overline{n}(P\overline{m}))))$.

Such a combinator exists by the fixed point theorem.

(2) The multiplication operation $\times$ is uniquely determined by the following two conditions:

    a. $n \times 0 = 0$.

    b. $n \times m^+ = (n \times m) + n$.

Thus the combinator $\otimes$ that we want, and which exists by the fixed point theorem, is defined by the condition:
$$\otimes\overline{n}\,\overline{m} = Z\overline{m}\overline{0}(\oplus(\otimes\overline{n}(P\overline{m}))\overline{n}).$$

(3) The exponential operation is uniquely determined by the following two conditions:

    a. $n^0 = 1$.

    b. $n^{m^+} = n^m \times n$.

Thus the combinator ⓔ we want, and which exists by the fixed point theorem, is defined by the condition $ⓔ\overline{m}\,\overline{n} = Z\overline{m}\overline{1}(\otimes(ⓔ\overline{n}(P\overline{m}))\overline{n})$.

8. By virtue of conditions (1) and (2) of the hint, we want a combinator Γ to satisfy the condition:

$$\Gamma x = Zxt(N(\Gamma(Px))).$$

Such a combinator can be found by the fixed point theorem.

9. No, it does not follow that $N\Gamma$ computes $\overline{A}$. Suppose $n \in A$. Then $\Gamma\overline{n} = t$. However, it does not follow that $N\Gamma\overline{n} = f$; what *does* follow is that $N(\Gamma\overline{n}) = f$. Thus it is not $N\Gamma$ that computes $\overline{A}$ but rather $BN\Gamma$ $[Bxyz = x(yz)]$. For $BN\Gamma\overline{n} = N(\Gamma\overline{n}) = f$ when $\Gamma\overline{n} = t$ (i.e. when $n \in A$) and $BN\Gamma\overline{n} = N(\Gamma\overline{n}) = t$ when $\Gamma\overline{n} = f$ (i.e. when $n \in \overline{A}$).

10. We want a combinator g such that for all numbers x and y the following conditions hold:
    (1) If $Z\overline{x} = t$, then $g\overline{x}\,\overline{y} = f$.
    (2) If $Z\overline{x} = f$, then
        (a) If $Z\overline{y} = t$, then $g\overline{x}\,\overline{y} = t$.
        (b) If $Z\overline{y} = f$, then $g\overline{x}\,\overline{y} = g(P\overline{x})(P\overline{y})$.
    We thus want g to satisfy the following:

$$gxy = Zxf(Zyt(g(Px)(Py))).$$

Such a g exists by the fixed point theorem.

11. Suppose A_1 realizes $f(x)$ and A_2 realizes $g(x)$. Then the functions $f(g(x))$ is realized by BA_1A_2, since

$$BA_1A_2\overline{n} = A_1(A_2\overline{n}) = A_1\overline{g(n)} = \overline{f(g(n))}.$$

12. Let Γ compute the relation $R(x, y)$, and let A_1 realize $f(x)$ and A_2 realize $g(x)$. Then:
    (a) $B\Gamma A_1$ computes the relation $R(f(x), y)$, because:

$$B\Gamma A_1\overline{n}\,\overline{m} = \Gamma(A_1\overline{n})\overline{m} = \Gamma\overline{f(n)}\,\overline{m},$$

which is t if $R(f(n), m)$ holds, and is f if $R(f(n), m)$ doesn't hold. Thus $B\Gamma A_1$ computes $R(f(x), y)$.
    (b) This is trickier! A combinator that computes the relation $R(x, g(y))$ is $BCD\Gamma A_2$, where C is the cardinal $[Cxyz = xzy]$ and D is the combinator defined by $Dxyzw = xy(zw)$. We learned that $D = BB$.

Well,

$$BCD\Gamma A_2\overline{n}\,\overline{m} = C(D\Gamma)A_2\overline{n}\,\overline{m} = D\Gamma\overline{n}A_2\overline{m} = \Gamma\overline{n}(A_2\overline{m}) = \Gamma\overline{n}\,\overline{g(m)},$$

which is t if $R(n, g(m))$ holds, and is f if $R(n, g(m))$ doesn't hold. Thus $BCD\Gamma A_2$ computes $R(x, g(y))$.

(c) Let $S(x, y)$ be the relation $R(x, g(y))$, which is computed by $BCD\Gamma A_2$. Then by (a) $S(fx), y)$, which is $R(f(x), g(y))$ is computed by $B(BCD\Gamma A_2)A_1$.

Alternatively, let $\delta(x, y)$ be the relation $R(f(x), y)$. It is computed by $B\Gamma A_1$, according to (a). Then $R(f(x), g(y))$ is $\delta(x, g(y))$, hence is computed by $BCD(B\Gamma A_1)A_2$, according to (b).

13. Given a regular relational combinator A of degree 2, by the fixed point theorem, there is a combinator A_1 satisfying the condition:

$$A_1 xy = (Axy)y(A_1 x(\sigma y)).$$

Thus, for any numbers $\overline{n}$ and $\overline{m}$, $A_1 \overline{n}\,\overline{m} = (A\overline{n}\,\overline{m})\overline{m}(A_1 \overline{nm + 1})$.

If $A\overline{n}\,\overline{m} = t$ then $A_1 \overline{n}\,\overline{m} = t\overline{m}(A_1 \overline{nm + 1}) = \overline{m}$.

If $A\overline{n}\,\overline{m} = f$ then $A_1 \overline{n}\,\overline{m} = f\overline{m}(A_1 \overline{nm + 1}) = A_1 \overline{nm + 1}$.

14. We take A' to be $CA_1\overline{0}$, where C is the cardinal and A_1 is the combinator of Problem 13. Then, for any number n, $A'\overline{n} = CA_1\overline{0}\overline{n} = A_1\overline{n}\overline{0}$. Now let k be the smallest number such that $A\overline{n}\overline{k} = t$. We illustrate the proof for $k = 3$ (the reader should have no trouble in generalizing the proof for arbitrary k). Thus it must be the case that $A\overline{n}\overline{0} = f$; $A\overline{n}\overline{1} = f$; $A\overline{n}\overline{2} = f$; $A\overline{n}\overline{3} = t$.

Now, $A'\overline{n} = A_1\overline{n}\overline{0}$. Since $A\overline{n}\overline{0} = f$, then $A_1\overline{n}\overline{0} = A_1\overline{n}\overline{1}$. Since $A\overline{n}\overline{1} = f$, then $A_1\overline{n}\overline{1} = A_1\overline{n}\overline{2}$. Since $A\overline{n}\overline{2} = f$, then $A_1\overline{n}\overline{2} = A_1\overline{n}\overline{3}$. Since $A\overline{n}\overline{3} = t$, then $A_1\overline{n}\overline{3} = \overline{3}$. Thus $A'\overline{n} =, A_1\overline{n}\overline{0} = A_1\overline{n}\overline{1} = A_1\overline{n}\overline{2} = A_1\overline{n}\overline{3} = \overline{3}$. So $A'\overline{n} = \overline{3}$.

15. Let $R(x, y)$ be the relation $10^y > x$. We must first find a combinator A that computes the relation $10^y > x$.

We let $R_1(x, y)$ be the relation $R(y, x)$, i.e. the relation $10^x > y$. If we find a combinator A_1 that computes the relation $R_1(x, y)$, then CA_1 will compute the relation $R(x, y)$ (why?).

To compute the relation $10^x > y$, we know that g computes the relation $x > y$ and that Ⓔ $\overline{10}$ realizes the function 10^x. Hence by Problem 11, the relation $10^x > y$ is computed by Bg(Ⓔ $\overline{10}$). Thus $C(Bg$(Ⓔ $\overline{10}$)) computes the relation $10^y > x$. We thus take the length measurer L' to be a minimizer of $C(Bg$(Ⓔ $\overline{10}$)).

16. Take ⊛ such that $⊛xy = \oplus(\otimes x$(Ⓔ $\overline{10}(L'y))y$.

17. Let A be a combinator satisfying the equation $A\overline{n} = \overline{3 * s * n * 4}$. [Specifically, we can take A to be $B(C\circledast \overline{4})(\circledast \overline{3 * s})$, where B is the bluebird and C is the cardinal, as the reader can verify.] Then $A\overline{n^{\#}} = \overline{(n+1)^{\#}}$ [since $A\overline{n^{\#}} = \overline{3 * s * n^{\#} * 4} = \overline{(n+1)^{\#}}$].

Then, by the Recursion Property, or more directly by Problem 6, there is a combinator δ such that:

(1) $\delta\overline{0} = \overline{0^{\#}}$.

(2) $\delta\overline{n+1} = A(\delta\overline{n})$.

By mathematical induction it follows that $\delta\overline{n} = \overline{n^{\#}}$, by the following reasoning:

(a) $\delta\overline{0} = \overline{0^{\#}}$, by (1).

(b) Now suppose n is such that $\delta\overline{n} = \overline{n^{\#}}$.

We must show that $\delta\overline{n+1} = \overline{(n+1)^{\#}}$. Well, by (2), $\delta\overline{n+1} = A(\delta\overline{n})$. And, of course, $\delta\overline{n} = \overline{n^{\#}}$, by the induction hypothesis. Therefore,

$$A(\delta\overline{n}) = A\overline{n^{\#}} = \overline{(n+1)^{\#}}.$$

And so

$$\delta\overline{n+1} = \overline{(n+1)^{\#}}.$$

This completes the induction.

18. Take Δ to be $S \circledast \delta$, where S is the starling $[Sxyz = xz(yz)]$. Then, recalling that $\delta\overline{n} = \overline{n^{\#}}$ we have:

$$\Delta\overline{n} = S \circledast \delta\overline{n} = \circledast\,\overline{n}(\delta\overline{n}) = \circledast\,\overline{n}\,\overline{n^{\#}} = \overline{n * n^{\#}}.$$

Note: In *To Mock a Mockingbird*, I took Δ to be $W(DC\circledast\delta)$, which also works, which should not be surprising, since for any x, y and z, $Sxyz = W(DCxy)z$, as the reader can verify.

19. It is (b) that is true. Let n be the Gödel number of X. Then $\overline{n} = \lceil X \rceil$, so that $\Delta\overline{n} = \Delta\lceil X \rceil$, and since $\Delta\overline{n} = \overline{n * n^{\#}}$, we have:

(1) $\Delta\lceil X \rceil = \overline{n * n^{\#}}$.

Since n is the Gödel number of X, then $n^{\#}$, which we first defined to be the Gödel number of $\overline{n}$, must also be the Gödel number of $\lceil X \rceil$, since $\overline{n} = \lceil X \rceil$. Thus $n * n^{\#}$ is the Gödel number of $X\lceil X \rceil$, and so:

(2) $\overline{n * n^{\#}} = \lceil X\lceil X \rceil \rceil$

(c) By combining (1) and (2), we see (b), i.e. that $\Delta\lceil X \rceil = \lceil X\lceil X \rceil \rceil$.

20. We take X to be $BA\Delta\lceil BA\Delta\rceil$ where Δ is the normalizer. By the definition of the bluebird B, we have $BA\Delta\lceil BA\Delta\rceil = A(\Delta\lceil BA\Delta\rceil)$. Consequently,
    (1) $X = A(\Delta\lceil BA\Delta\rceil)$ But also:
    (2) $A(\Delta\lceil BA\Delta\rceil) = A\lceil X\rceil$.
    Here is the reason for (2): By Problem 19, $\Delta\lceil Y\rceil = \lceil Y\lceil Y\rceil\rceil$ for any Y. Hence
    $$\Delta\lceil BA\Delta\rceil = \lceil BA\Delta\lceil BA\Delta\rceil\rceil,$$
    which is $\lceil X\rceil$. Since $\Delta\lceil BA\Delta\rceil = \lceil X\rceil$, then $A(\Delta\lceil BA\Delta\rceil) = A\lceil X\rceil$, which proves (2).
        By (1) and (2), we have $X = A\lceil X\rceil$ and so $A\lceil X\rceil = X$.

21. Suppose the function $f(x)$ is realizable, and let A realize $f(x)$. By the second fixed point principle just proved, there is a term X such that $X = A\lceil X\rceil$. Let n be the Gödel number of X, so that $\bar{n} = \lceil X\rceil$, and therefore $A\lceil X\rceil = A\bar{n}$ and so $X = A\bar{n}$. Also $A\bar{n} = \overline{f(n)}$, since A realizes $f(x)$, which means that X designates $f(n)$, where n is the Gödel number of X.

22. Both (1) and (2) are true: Suppose Γ computes A. This means that for every n:
    (a) $n \in A$ implies that $\Gamma\bar{n} = t$;
    (b) $n \notin A$ implies that $\Gamma\bar{n} = f$.

    (1) We must show that A is representable. Well, it is Γ itself that represents A. To see this, it suffices to see that the converse of (a) holds, i.e. that if $\Gamma\bar{n} = t$, then $n \in A$. So suppose $\Gamma n = t$. If n were not in A, then by (b) we would have that $\Gamma\bar{n} = f$, and hence that $t = f$, which cannot be. Thus $n \in A$.
    (2) Now we must show that $\overline{A}$ is representable. Since Γ computes A, then $BN\Gamma$ computes the complement $\overline{A}$ of A [by the solution of Problem 9]. Thus $BN\Gamma$ represents $\overline{A}$ [by (1)].

23. Suppose that Γ_1 realizes $f(x)$ and that Γ_2 represents A. We let $\Gamma = B\Gamma_2\Gamma_1$ $[Bxyz = x(yz)]$, and we show that Γ represents $f^{-1}(A)$. Well,
    $$\Gamma\bar{n} = B\Gamma_2\Gamma_1\bar{n} = \Gamma_2(\Gamma_1\bar{n}) = \Gamma_2\overline{f(n)}.$$
    Thus $\Gamma\bar{n} = \Gamma_2\overline{f(n)}$, so that $\Gamma\bar{n} = t$ iff $\Gamma_2\overline{f(n)} = t$, which is true iff $f(n) \in A$ [since Γ_2 represents A], which is true iff $n \in f^{-1}(A)$. Thus $\Gamma\bar{n} = t$ iff $n \in f^{-1}(A)$, which means that Γ represents $f^{-1}(A)$.

24. This is but a special case of Problem 23: Let $f(x) = x * 52$. By Problem 16, $f(x)$ is realizable by the combinator $\circledast \overline{x52}$. If A is representable, so is $f^{-1}(A)$ [by Problem 23], and $f^{-1}(A)$ is the set of all n such that $f(n) \in A$, i.e. the set of all n such that $(n * 52) \in A$.

25. Suppose A is representable. Let A' be the set of all n such that the number $n * 52$ is in A. Then A' is representable [by Problem 24]. Let Γ represent the set A'. By the fixed point theorem, there is some X such that $\Gamma\lceil X\rceil = X$. We will show that the sentence $X = t$ is a Gödel sentence for A.

   Let n be the Gödel number of X. Then $\overline{n} = \lceil X\rceil$. Hence $\Gamma\overline{n} = \Gamma\lceil X\rceil$, and therefore $\Gamma\lceil X\rceil = t$ iff $\Gamma\overline{n} = t$, which is true iff $n \in A'$ [since Γ represents A']. And $n \in A'$ iff $(n * 52) \in A$. Thus $\Gamma\lceil X\rceil = t$ iff $(n * 52) \in A$. But $n * 52$ is the Gödel number of the sentence $X = t$. Thus the sentence $X = t$ is a Gödel sentence for A.

26. There cannot be a Gödel sentence for the *complement* of T_0, for such a sentence X would be true if and only if its Gödel number n was not in T_0, which means that X would be true iff its Gödel number was not the Gödel number of a true sentence, and this is impossible. [Looked at it another way, every sentence is a Gödel sentence for the set T_0, and no sentence can be a Gödel sentence for both a set A and its complement (why?). Hence no sentence can be a Gödel sentence for the complement of T_0.]

   Since there is no Gödel sentence for the complement of T_0, then the complement of T_0 is not representable [by Problem 25], and therefore the set T_0 is not computable [by Problem 22].

Afterword
Where to Go from Here

This book and its predecessor have only scratched the surface of the collection of topics subsumed under the name of "Mathematical Logic." We have done only the beginnings of the fields of recursion theory and combinatory logic, and have done virtually nothing in model theory, proof theory, and other logics such as modal logic, intuitionistic logic, relevance logic, and others.

I strongly suggest that you next turn your attention to set theory and the continuum problem. Let me tell you a wee bit about this, hoping to whet your appetite.

One purpose of the subject known as *axiomatic set theory* is to develop all mathematics out of the notions of logic (the logical connectives and the quantifiers) together with the notion of an element being a *member* of a set of elements. We have used the symbol "$\in$" for "is a member of" and "$x \in A$" for "x is a member of the set A".

A pioneer in the development of axiomatic set theory was Gottlob Frege [1893]. His system had, in addition to axioms of first-order logic, just one axiom of set theory, namely that for any property P, there is a unique set consisting of those and only those things that have the property P. Such a set is written $\{x : P(x)\}$, which is read "the set of all x's having the property P.

This principle of Frege is sometimes referred to as the *abstraction principle*, or the *unlimited abstraction principle*. One can define the identity relation $x = y$ in terms of set inclusion as "x and y belong to the same sets" [i.e. $\forall z(x \in z \equiv y \in z)$]. Frege's abstraction principle has the marvelous advantage of allowing us to obtain just about all the sets necessary for mathematics, for example, the following:

P_1: The empty set $\emptyset$, which is $\{x : \sim (x = x)\}$.

P_2: For any two elements a and b, the set $\{a, b\}$ whose members are just a and b. This set is $\{x : x = a \vee x = b\}$.

P_3: For any set a, the power set $\mathcal{P}(a)$, i.e. $\{y : y \subseteq a\}$, thus

$$\{y : \forall z (z \in y \supset z \in a)\}.$$

P_4: For any set a, the union $\cup\, a$, i.e. the set of all elements that are members of an element of a, thus $\cup\, a = \{x : \exists y (y \in a \wedge x \in y)\}$.

Despite the useful things Frege's system can do, it has an extremely serious drawback: it is inconsistent! This was pointed out to Frege in a letter by Bertrand Russell [1902], who observed that according to Frege's abstraction principle, there would exist the set A of all sets which are not members of themselves ($A = \{x : x \notin x\}$). Thus, for any set x, one would have $x \in A$ iff $x \notin x$. Taking A for x, we would have $A \in A$ iff $A \notin A$, which is a contradiction! Thus Frege's system is inconsistent.

Frege was broken-hearted over Russell's discovery, and felt that his whole life's work had been in vain. Here he was wrong: His work was salvaged by Zermelo and others, and is really the basis for Zermelo–Fraenkel set theory (frequently abbreviated as ZF), which is one of the most significant systems of set theorem in existence, the other of similar rank being the system in *Principia Mathematica* of Whitehead and Russell [1910].

What Zermelo [1908] did was to replace Frege's abstraction principle by the following, known as the *limited abstraction principle*, which is that for any property P of sets *and any set a*, there exists the set of *all the elements of a* having property P. Thus for any set a, there exists the set $\{x : P(x) \wedge x \in a\}$. This principle appears to be free of any possible contradiction. However, Zermelo had to take the existence of the sets $\emptyset$, $\{a, b\}$, $\mathcal{P}(a)$, and $\cup\, a$ as separate axioms. He also took an additional axiom known as the *axiom of infinity*, which we will discuss later.

On the basis of the Zermelo axioms on hand, one can now derive the positive integers. First, let us note that the set $\{a, b\}$ is well-defined even if a and b are the same element, and then $\{a, a\}$ is simply the set $\{a\}$. Well, Zermelo took 0 to be the empty set $\emptyset$; 1 to be the set whose only element is 0; 2 to be the set whose only element is 1, and so forth. Thus the Zermelo natural numbers are $\emptyset$, $\{\emptyset\}$, $\{\{\emptyset\}\}$, etc. Thus for each natural number n, $n+1$ is $\{n\}$.

Later, Von Neumann [1923] proceeded differently. In his scheme, each natural number is the set of all lesser natural numbers. Thus 0 is $\emptyset$ and 1 is $\{\emptyset\}$ as with Zermelo, but 2 is now not $\{1\}$, but rather the set $\{0,1\}$; 3 is the set $\{0,1,2\}$; $\ldots, n+1$ is the set $\{0,1,\ldots,n\}$. This is the scheme now generally adopted, since it generalizes in the infinite case to the important sets called *ordinals*, which we will soon discuss. Note that in Zermelo's scheme, each natural number other than 0 contains just one element, whereas in Von Neumann's scheme, each number n contains exactly n elements.

Now that we have the natural numbers, how do we define what it means to be a natural number? That is, we want a formula $N(x)$ using just the logical connectives and quantifiers and the single predicate symbol $\in$ such that for any set a, the sentence $N(a)$ is true if and only if a is a natural number. Well, for any set a, we define a^+ as $a \cup \{a\}$, i.e. the set whose elements are those of a, together with a itself. Thus

$$1 = 0^+ = \{0\}, \quad 2 = 1^+ = 1 \cup \{1\}; \quad 3 = 2^+ = 2 \cup \{2\}$$

(which is $\{0,1,2\}$), etc.

We now define a set a to be *inductive* if $0 \in a$, and if, for all sets b, if $b \in a$, then $b^+ \in a$. We then define a set x to be a *natural number* if it belongs to all inductive sets. The principle of mathematical induction follows immediately from the very definition of a natural number. Indeed, the five Peano Postulates follow from the definition of "natural number".

However, we don't yet have any guarantee that there is such a thing as the *set* of all natural numbers. Zermelo had to take this as an additional axiom, and this is known as the *axiom of infinity*. The set of all natural numbers is denoted "ω".

Let us recall the Peano Postulates:

1. 0 is a natural number.
2. If n is a natural number, so is n^+.
3. There is no natural number such that $n^+ = 0$.
4. If $n^+ = m^+$, then $n = m$.
5. [Principle of Mathematical Induction] Every inductive set contains all the natural numbers.

All these five postulates are easy consequences of the definition of natural number. The axiom of infinity is *not* necessary to prove the Peano Postulates! This fact appears to not be completely well known. I recall that when I once announced these facts at a lecture, a very eminent logician later told me that he was quite shocked!

Now for the ordinals: First we define a set x to be *transitive* if it contains with each of its elements y all elements of y as well; in other words, each element of x is a subset of x. We recall that for any set a, we are taking a^+ to be the set $a \cup \{a\}$, and a^+ is called the *successor of a*. Roughly speaking, the ordinals are those sets, starting with $\emptyset$, which can be obtained by taking the successor of any ordinal already obtained, and by taking any transitive set of ordinals already obtained. That is, we want to define *ordinals* such that 0 is an ordinal, every successor of an ordinal is an ordinal, and every transitive set of ordinals is an ordinal. One definition that works is this: Define x to be an ordinal if x is transitive and every transitive *proper* subset of x is a member of x. [We recall that by a *proper* subset of x is meant a subset of x that is not the whole of x.] It can be seen from this definition that $\emptyset$ is an ordinal, the successor of any ordinal is an ordinal, and any transitive set of ordinals is an ordinal.

It is obvious that every natural number is an ordinal (use mathematical induction). Also, the set ω of all natural numbers is transitive (verify!), and so ω is an ordinal. Once we have ω, we have the ordinal ω^+, also denoted $\omega + 1$. Successively, we have $\omega + 2$ $(\omega^{++}), \omega + 3, \ldots, \omega + n, \ldots$ However, we do not yet have a set which contains all these ordinals, and the existence of such a set cannot be derived from Zermelo's axioms. But then Abraham Fraenkel [1958] added another axiom known as the *axiom of substitution*, also called the *axiom of replacement*, which does ensure the existence of such a set, and this set is denoted $\omega \times 2$ or $\omega \cdot 2$.

Roughly speaking, the axiom of replacement is that given any operation F which assigns to each set x a set $F(x)$, and given any set a, there is a set [denoted $F^{-1}(a)$] consisting of all, and only, those sets x such that $F(x) \in a$.

Well, there is the set denoted $\omega \cdot 2$ of all elements $\omega + n$, where n is any natural number. Then we have the ordinals

$$\omega \cdot 2 + 1, \quad \omega \cdot 2 + 2, \quad \omega \cdot 2 + 3, \ldots, \omega \cdot 2 + n, \ldots, \omega \cdot 3.$$

After that finally come $\omega \cdot 4$, $\omega \cdot 5$, ..., $\omega \cdot \omega$, ... ω^ω, ... We can keep going.

The ordinals so far considered are all denumerable. But there exist non-denumerable ordinals as well. Indeed, with the axiom of substitution/replacement (of Zermelo–Fraenkel set theory), for every set x, there is an ordinal α of the same size as x (i.e. α can be put into 1-1 correspondence with x).

An ordinal is called a *successor ordinal* if it is $\alpha^+[\alpha \cup \{\alpha\}]$ for some ordinal α. Ordinals other than 0 and successor ordinals are called *limit ordinals*. All natural numbers other than 0 are successor ordinals. The first limit ordinal is ω. The ordinals $\omega + 1, \omega + 2, \ldots, \omega + n$ are all successor ordinals, and $\omega \cdot 2$ (the set of all of them) is the next limit ordinal. The ordinals $\omega \cdot 3, \omega \cdot 4, \ldots, \omega \cdot n$ are all limit ordinals, and so is $\omega \cdot \omega$. Actually, an infinite ordinal is a successor ordinal if and only if it is of the form $\alpha + n$, for some limit ordinal α and some *positive* integer n.

An ordinal α is said to be *less than* an ordinal β (in symbols, $\alpha < \beta$) if $\alpha \in \beta$. Thus each ordinal is the collection of all lesser ordinals.

Rank

Given a scheme that assigns to each ordinal α a set denoted S_α, for any ordinal λ, by $U_{\alpha<\lambda}S_\alpha$ is meant the union of all the sets S_α, where $\alpha < \lambda$. By an important result known as the *transfinite recursion theorem*, one can assign to each ordinal α a set R_α such that for every ordinal α and every *limit ordinal* λ, the following three conditions hold:

(1) $R_0 = \emptyset$.
(2) $R_{\alpha^+} = \mathcal{P}(R_\alpha)$ [the set of all subsets of R_α].
(3) $R_\lambda = U_{\alpha<\lambda}R_\alpha$.

A set is said to have *rank* if it is a member of some R_α, and by the *rank* of such a set x is meant the least ordinal α such that $x \in R_\alpha$.

Do all sets have rank? This is so iff the $\in$ relation is *well founded*, i.e. iff there exists no infinite sequence $x_1, x_2, \ldots x_n, \ldots$ such that for each n, the set x_{n+1} is a member of x_n. This condition is taken as another axiom of set theory, and is known as the *axiom of foundation*. It rules out sets without rank.

The Axiom of Choice and the Generalized Continuum Hypothesis

The Axiom of Choice (frequently abbreviated AC) in one form is that for any non-empty set a of non-empty sets, there is a function C (a so-called *choice function* for a) that assigns to each member x of a an element of x. [The function, so to speak, *chooses* one element from each of the elements of a.]

An equivalent form of AC is that for any non-empty set a of non-empty sets, there is a set b whose members consist of just one member of each of the members of a.

What is the status of AC? Most working mathematicians of the world accept it as being true. There are a few, though, who don't. What is its status with respect to ZF (Zermelo–Fraenkel Set Theory)? Well, Gödel [1940] showed that it is *consistent* with ZF (assuming that ZF is itself consistent, which we will continue to assume). Thus AC is not refutable in ZF. Sometime later, Paul Cohen [1963, 1964] proved that the *negation* of AC is consistent with ZF. Thus AC is undecidable in ZF.

We recall Cantor's theorem, that for any set a, its power set $\mathcal{P}(a)$ is larger than a. Cantor conjectured that for an *infinite* set there is no set b intermediate in size between a and $\mathcal{P}(a)$ [that is, that there is no set b which is larger than a but smaller than $\mathcal{P}(a)$], and this result is known as the *Generalized Continuum Hypothesis* (frequently abbreviated as GCH). What is its status? This is far more puzzling! Unlike the case of the axiom of choice, which most logicians believe is true, most mathematicians, as well as most logicians specializing in set theory, don't have the slightest idea as to whether GCH is true or false. Gödel *conjectured* that it is false, despite the fact that he proved it consistent with the axioms of ZF. His proof of the consistency of GCH with ZF is surely one of the most remarkable things in mathematical logic! Equally remarkable is the proof of Paul Cohen that the *negation* of GCH is consistent with ZF, even when AC is added to the axioms of ZF. Thus GCH is undecidable in ZF.

Gödel's proof of the consistency of AC and GCH used his notion of *constructible sets*, to which we now turn.

Constructible Sets

Consider a sentence X whose constants are all in a set a. X is said to be *true over* a if it is true when the quantifiers are interpreted as ranging over all elements of a, i.e. it is true when "$\forall x$" is read as "for all x in a" and "$\exists x$" is read as "for some x in a". A formula $\varphi(x)$ whose constants are all in a is said to *define over* a the set of all elements k such that $\varphi(k)$ is true over a, and a subset b of a is called *definable over* a (more completely, *first-order definable over* a) if it is defined over a by some formula whose constants are all in a.

We let $\mathcal{F}(a)$ be the set of all sets definable over a.

Gödel introduced the notion of *constructible sets* as follows: Again by the transfinite recursion theorem, one can assign to each ordinal α a set M_α such that for every ordinal α and every limit ordinal λ, the following three conditions hold:

(1) $M_0 = \emptyset$.
(2) $M_{\alpha^+} = F(M_\alpha)$.
(3) $M_\lambda = \bigcup_{\alpha<\lambda} M_\alpha$.

The definition of the M_α's differs from that of the R_α's only in condition (2): Whereas $R_{\alpha+1}$ consists of *all* subsets of R_α, the set $M_{\alpha+1}$ consists of only those subsets of M_α that are definable over M_α.

A set is called *constructible* if it is a member of some M_α.

Are all sets constructible? This is a grand unsolved problem! Gödel conjectured that this is not the case, even though he proved that it was consistent with the axioms of ZF. The significance of the constructible sets is that the constructability of all sets implies the generalized continuum hypothesis! Moreover, this implication is provable in ZF. Thus if we add to the axioms of ZF the axiom that all sets are constructible (which is known as the axiom of constructability), the continuum hypothesis (as well as the axiom of choice) becomes provable. And since, as Gödel showed, the axiom of constructability is consistent with ZF, so is the generalized continuum hypothesis (as well as the axiom of choice). This is how Gödel showed the consistency of GCH with ZF.

Some years later, Paul Cohen [1963, 1964] showed, by a completely different method, that the negation of GCH is consistent with ZF, even with the addition of AC. Thus GCH is undecidable in ZF.

Despite the undecidability of GCH in ZF, the question remains as to whether or not it is really true. To many so-called "formalists", the question has no meaning. They say that is provable in some axiom systems and disprovable in others. To the so-called "Platonist" like Gödel, this position is most unsatisfactory. I would liken it to the following: Suppose a bridge is built and on the next day an army is to march over it. Will the bridge hold or not? It does no good to say, "Well, in some axiom systems, it can be proved that the bridge will hold, and in others, it can be proved that it won't." We want to know whether it will *really* hold or not. And likewise with the generalized continuum hypothesis. It is not enough to say it is undecidable in ZF. The fact remains that for every infinite set a, either there is a set b intermediate in size between a and $\mathcal{P}(a)$ or there isn't, and we want to know which. As already, indicated, Gödel conjectured that there is such a set b for some infinite set a, and that when more is known about sets, we will see that Cantor's conjecture is wrong. [Realize that for any non-empty *finite* set a, there are always sets b with more elements than a and less elements than $\mathcal{P}(a)$.]

The whole study of the independence of GCH in ZF is fascinating beyond belief, and the reader who studies this is in for a real treat!

References

Barendregt, Henk, *The Lambda Calculus — Its Syntax and Semantics*, Studies in Logic and the Foundations of Mathematics Vol. 103, North-Holland, 1985.

Beth, Evert, *The Foundations of Mathematics. A Study in the Philosophy of Science*, North-Holland, 1959.

Braithwaite, Reginald, *Kestrels, Quirky Birds and Hopeless Egocentricity*, ebook published by http://leanpub.com, 2013.

Church, Alonzo, *The Calculi of Lambda Conversion*, Princeton Univ. Press, 1941.

Cohen, Paul J., "The independence of the continuum hypothesis," *Proceedings of the National Academy of Sciences of the United States of America* **50**(6): 1143–1148, 1963.

Cohen, Paul J., "The independence of the continuum hypothesis, II," *Proceedings of the National Academy of Sciences of the United States of America* **51**(1): 105–110, 1964.

Craig, William, "Three uses of the Herbrand–Gentzen theorem in relating model theory and proof theory," *The Journal of Symbolic Logic* **22**(3): 269–285, 1957.

Davis, Martin, *Computability and Unsolvability*, McGraw-Hill, 1958; Dover, 1982.

Ehrenfeucht, Andrzej and Feferman, Solomon, "Representability of recursively enumerable sets in formal theories," *Arch. Fur Math. Logick und Grundlagenforschung* **5**: 37–41, 1960.

Fraenkel, Abraham Bar-Hillel, Yehoshua and Levy Azriel, *Foundations of Set Theory*, North-Holland, 1973 (originally published in 1958). (Fraenkel's final word on ZF and ZFC, according to Wikipedia.)

Frege, Gottlob, *Grundgesetze der Arithmetic*, Verlag Hermann Pohle, Vol. I/II, 1893. Partial translation of volume I, *The Basic Laws of Arithmetic*, by M. Furth, Univ. of California Press, 1964.

Gentzen, Gerhard, "Untersuchungen über das logische Schliessen I," *Mathematische Zeitschrift* **39**(2): 176–210, 1934.

Gentzen, Gerhard, "Untersuchungen über das logische Schliessen II," *Mathematische Zeitschrift* **39**(3): 405–431, 1935.

Gödel, Kurt, *The Consistency of the Axiom of Choice and of the Generalized Continuum Hypothesis with the Axioms of Set Theory*, Princeton Univ. Press, 1940.

Henkin, Leon, "The completeness of the first-order functional calculus," *J. Symbolic Logic*, **14**: 159–166, 1949.

Hintikka, Jaakko, "Form and content in quantification theory," *Acta Philosophica Fennica*, **8**: 7–55. 1955.

Kleene, Stephen Cole, "Recursive predicates and quantifiers," *Trans. Amer. Math. Soc.* **53**: 41–73, 1943.

Kleene, Stephen Cole, *Introduction to Metamathematics*, North-Holland, 1952.

Myhill, John, "Creative sets," *Z. Math. Logik Grundlagen Math.* **1**: 97–108, 1955.

Post, Emil Leon, "Formal reductions of the general combinatorial decision problem," *American Journal of Mathematics* **65**: 197–215, 1943.

Post, Emil Leon, "Recursively enumerable sets of positive integers and their decision problems," *Bull. Amer. Math. Soc.* **50**(5): 284–316, 1944.

Putnam, Hilary and Smullyan, Raymond, "Exact separation of recursively enumerable sets within theories," *Journal of the American Mathematical Society*, **11**(4): 574–577, 1960.

Rice, H. Gordon, "Classes of recursively enumerable sets and their decision problems," *Trans. Amer. Math. Soc.* **74**(2): 358–366, 1953.

Rosser, John Barkley, "Extensions of some theorems of Gödel and Church," *Journal of Symbolic Logic*, **1**(3): 87–91, 1936.

Russell, Bertrand, "Letter to Frege," 1902. This very interesting letter can be found (translated from the German) in Van Heigenoort, Jean, *From Frege to Gödel, A Source Book in Mathematical Logic, 1879–1931*, Harvard Univ. Press, 1967. It is also available online at a Harvard website: http://isites.harvard.edu/fs/docs/icb.topic1219929.files/FregeRussellCorr.pdf

Schönfinkel, Moses, "Über die Bausteine der mathematischen Logik", *Mathematische Annalen* **92**, 1924; translated as "On the building blocks of mathematical logic" and included in Van Heigenoort, Jean, *From Frege to Gödel, A Source Book in Mathematical Logic, 1879–1931*, Harvard Univ. Press, 1967.

Shepherdson, John, "Representability of recursively enumerable sets in formal theories," *Archiv für Mathematische Logik und Grundlagenforschung*, 119–127, 1961.

Smullyan, Raymond, *Theory of Formal Systems*, Princeton Univ. Press, 1961.

Smullyan, Raymond, "A unifying principle in quantification theory", *Proceedings of the National Academy of Sciences*, **49**(6): 828–832, 1963.

Smullyan, Raymond, *To Mock a Mockingbird*, Alfred A. Knopf, 1985.

Smullyan, Raymond, *Gödel's Incompleteness Theorems*, Oxford Univ. Press, 1992.

Smullyan, Raymond, *Recursion Theory for Metamathematics*, Oxford, 1993.

Smullyan, Raymond, *Diagonalization and Self-Reference*, Oxford Science Publications, 1994.

Smullyan, Raymond, *First-Order Logic*, Springer-Verlag, 1968; Dover, 1995.

Smullyan, Raymond, *Logical Labyrinths*, CRC Press, 2008; A. K. Peters, Ltd. 2009.

Smullyan, Raymond, *The Beginner's Guide to Mathematical Logic*, Dover, 2014.

Sprenger, M. and Wymann-Böni, M., "How to decide the lark," *Theoretical Computer Science*, **110**: 419–432, 1993.

Statman, Richard, "The word problem for Smullyan's lark combinator is decidable," *Journal of Symbolic Computation*, **7**(2): 103–112, 1989.

Tarski, Alfred, *Undecidable Theories*, North-Holland, 1953.

Von Neumann, John, "On the introduction of transfinite numbers," in Jean van Heijenoort, *From Frege to Gödel: A Source Book in Mathematical Logic, 1879–1931* (3rd ed.), Harvard Univ. Press, 1923, pp. 346–354 (English translation of von Neumann 1923), 1973.

Whitehead, Alfred North and Russell, Bertrand, *Principia Mathematica*, Cambridge Univ. Press, 1910, Vol. 1. Reprinted by Rough Draft Printing, 2011.

Zermelo, Ernst, "Untersuchungen über die Grundlagen der Mengenlehre I," *Mathematische Annalen* **65**(2): 261–281, 1908. English translation: Heijenoort, Jean van (1967), "Investigations in the foundations of set theory," *From Frege to Gödel: A Source Book in Mathematical Logic, 1879–1931*, Source Books in the History of the Sciences, Harvard Univ. Press, 1967, pp. 199–215.

Index

1-1 (one-to-one) function, 97

$\delta(x, y)$, recursive pairing function, 97

$\delta_n(x_1, \ldots, x_n)$, recursive n-tupling function, 98

Γ-consistent set of sentences, 49–56

λ-I combinators, 192–196

ω_n, 113

$\Phi, \Phi_2, \Phi_3, \Phi_4$ combinators, 195, 196

Φ_n combinator, 196

σ, successor combinator, 234

Σ_0 formula of elementary arithmetic, 94, 95

Σ_0 set or relation, 95

Σ_1 formula of elementary arithmetic, 95

Σ_1 set or relation, 93, 95

$\sim AC$ is consistent with ZF (Cohen), 258

A^*, C^*, R^*, F^* and V^* combinators, 189

A^* number set in a simple system, 151

$A^{**}, C^{**}, R^{**}, F^{**}$ and V^{**} combinators, 189

abstraction principle, 253

AC (Axiom of Choice), 258–260

AC is consistent with ZF (Gödel), 258

AC is undecidable in ZF (Gödel and Cohen), 258

addition combinator, 234, 235

admissible function in a simple system, 152

affirmation of a number by a register, 70

affirmation set of a register, 73

agreeable element, in combinatory logic, 175

algebraic approach to verifying tautologies, 4–7

alphabet of an elementary formal system, 89, 90

altered tableau method, 40–43

analytic consistency property, 49–56

analytic tableau, 34–36, 40, 49–56

applications of magic sets, 27, 28

applications of Rice's Theorem, 120

applications of simple systems, 150

arithmetic combinators, 234, 235

arithmetic set or relation, 95

arithmetic term of combinatory logic, 241

arithmetical combinator of type n, 234

associate regular set for a finite set of sentences, 53

atomic formula of an elementary formal system, 89, 90

Axiom of Choice (AC), 258–260

axiom of constructability, 259

axiom of foundation, 257

axiom of infinity, 254, 256

Printed in the United States
By Bookmasters